UNDERWATER EDEN

The New England Aquarium is redefining what it means to be an aquarium: combining education, entertainment, and action to address the most challenging problems facing the oceans. We are a global leader in ocean exploration and marine conservation and are the only Boston-based cultural institution with a mission focused primarily on the environment.

Building upon a strong foundation of science, partnership, and field demonstration, Conservation International empowers societies to responsibly and sustainably care for nature, our global biodiversity, for the well-being of humanity.

UNDERWATER EDEN

Saving the Last Coral Wilderness on Earth

Edited by Gregory S. Stone & David Obura

THE UNIVERSITY OF CHICAGO PRESS   *Chicago and London*

The University of Chicago Press, Chicago 60637
The University of Chicago Press, Ltd., London
© 2013 by The New England Aquarium
All rights reserved. Published 2013.
Printed in China

22 21 20 19 18 17 16 15 14 13   1 2 3 4 5

ISBN-13: 978-0-226-77560-9 (cloth)
ISBN-10: 0-226-77560-7 (cloth)
ISBN-13: 978-0-226-92267-6 (e-book)
ISBN-10: 0-226-92267-7 (e-book)

Library of Congress Cataloging-in-Publication Data

Underwater Eden: saving the last coral wilderness on
earth / edited by Gregory S. Stone & David Obura.
    pages; cm.
  Includes bibliographical references and index.
  ISBN-13: 978-0-226-77560-9 (cloth: alkaline paper)
  ISBN-10: 0-226-77560-7 (cloth: alkaline paper)
  ISBN-13: 978-0-226-92267-6 (e-book)
  ISBN-10: 0-226-92267-7 (e-book) 1. Coral reef conser-
vation—Kiribati—Phoenix Islands. 2. Coral reefs and
islands—Kiribati—Phoenix Islands. 3. Phoenix Islands
(Kiribati)—Environmental conditions. I. Stone, Gregory S.
II. Obura, David O.
  QH77.P45U53 2013
    639.9'736099681—dc23   2012012025

♾ This paper meets the requirements of ANSI/NISO
Z39.48-1992 (Permanence of Paper).

CONTENTS

FOREWORD

HIS EXCELLENCY ANOTE TONG
TE BERETITENTI (PRESIDENT)
REPUBLIC OF KIRIBATI

Kiribati is a republic of water, and nowhere in our thirty-three islands and atolls are you ever more than a few meters from the sea. We are not a wealthy nation in financial terms, but the riches we do have—material, cultural, and spiritual—come from the Pacific Ocean. The waters around us are some of the most productive fishing grounds for tuna and reef fish in the world. Kiribati has its own small-scale tuna fishery, but over the decades, with the rise of multinational fishing fleets, we came to rely on the sale of lucrative fishing rights to provide the most basic services to our people.

In 2006 Kiribati made news around the world by announcing that we were closing off some of our best coral reefs and tuna fishing grounds to create a marine protected area (MPA), the ocean equivalent of a Yellowstone National Park. Once the MPA was established, we began to phase out commercial fishing from

His Excellency Anote Tong, President of Kiribati, with Gregory Stone. (Sterling Zumbrunn)

North
Pacific
Ocean

Canada

United States

Mexico

Hawaii

Northern
Line
Group

Gilbert Group

THE REPUBLIC OF KIRIBATI

Phoenix
Group

**Phoenix Islands
Protected Area**

Southern
Line
Group

South
Pacific
Ocean

Australia

New
Zealand

Phoenix Islands
Protected Area Boundary

Country of Kiribati International Boundary*

Scale varies from this perspective
*International Boundaries are estimated and are not authoritative.

(New England Aquarium)

11 percent of our exclusive economic zone, or EEZ—some 400,000 square kilometers.

At the center of the MPA were the eight islands of the Phoenix group, home to a few dozen people and millions of nesting seabirds. Offshore, the unspoiled reefs teemed with fish and other sea life at a level of abundance known nowhere else in the world. Kiribati proposed to create a trust to offset the loss of fishing income and to permanently preserve the Phoenix Islands, with their birds and corals and fishes.

We took this extraordinary step based on the word of a handful of scientists who had come to us five years earlier with news of a discovery in our waters— the nearly unspoiled reefs of the Phoenix Islands. "You have an extraordinary treasure," they told us, "and an extraordinary opportunity to preserve it." Gregory Stone, David Obura, and their colleagues came to the government of Kiribati to show us what no previous scientists ever had: the wonderful, vibrant biodiversity unique to our waters, and the urgent need to save it.

The opportunity to save the Phoenix Islands became a firm commitment and shared cause, one championed from modest offices in Tarawa to legal boardrooms in Boston, Massachusetts, and Washington, DC, from government offices in Australia and New Zealand to international conferences in Brazil.

The establishment of the Phoenix Islands Protected Area was by no means straightforward. There were deep reservations within Kiribati and in the wider financial and conservation communities. Our fisheries partners vigorously contested the loss of some of their most fertile fishing grounds. Our own citizens protested over the potential loss of much-needed revenue. Naysayers maintained that the plan simply couldn't succeed, that such a tiny nation with so few resources couldn't police and enforce a huge marine protected area. But the Phoenix Islands Protected Area was established, then enlarged. In 2010 the Phoenix Islands were inscribed on the World Heritage list by the United Nations Environmental and Scientific Committee (UNESCO), becoming Earth's largest and deepest World Heritage site.

The Phoenix Islands teach us that much can be accomplished with sufficient political will and commitment. The people of Kiribati face the real possibility of being displaced by climate change as our planet warms, sea levels rise, and our low-lying atolls leave us nowhere to go. Beginning in 2003 the government launched the Kiribati Adaptation Program to respond to climate change by capturing rainwater, planting mangroves, and taking other steps to reduce Kiribati's vulnerability to rising sea levels and other effects of climate change. Since 2007, in partnership with AusAID, the Kiribati Ministry of Education has funded the training of I-Kiribati nurses at Griffith University in Australia. By providing economic incentives for emigration and allowing young people time to establish I Kiribati communities in other countries, the Kiribati Australia Nursing Initiative and programs like it seek to facilitate the eventual migration of the people of Kiribati by 2050 with dignity and hope.

Yet the people of Kiribati recognize that they are still stewards of the ocean that now threatens their way of life. They see the value of saving the Phoenix Islands, a natural treasure that belongs not just to Kiribati, but to the world.

Shark over lettuce corals.
(Brian Skerry)

The vibrant smile of this
I-Kiribati dancer being
dressed for a performance
will disappear when the
music plays. Smiling during
a traditional I-Kiribati dance
is considered vulgar, as the
dancer is using storytelling
to pay tribute to the spirits
while displaying endurance,
skill, and beauty. (Cat Holloway)

This book tells the story of the underwater Eden that is the Phoenix Islands and of the race to save this wilderness against great odds. It tells of the islands' past and present, but their future is yet to be written. We all write that future every day, as stewards of our shared ocean. It has been observed that human blood has the same salinity as the ancient seas from which life on land first emerged millions of years ago. The Phoenix Islands awaken us to the ancient truth that the ocean is within every one of us, and we must manage it as one people.

Members of the 2002 expedition aboard a skiff, preparing to dive into the pristine waters of the Phoenix Islands. (Paul Nicklen)

1 IN SEARCH OF PARADISE

GREGORY S. STONE

July 2002, 2°50'S, 171°40'W/2.833°S 171.667°W

A raucous throng of sooty terns hovered over Kanton Island, calling out in high-pitched screeches. Beyond the low sandy atoll, the South Pacific seemed to stretch forever beneath tropical clouds topped by immense crowns of gold and red. It was 6:30 a.m., and biologists David Obura and Sangeeta Mangubhai, ship's doctor Mary Jane Adams, dive master Cat Holloway, and I adjusted our scuba gear as we sat on the pontoon of the gently rocking outboard skiff.

"This is definitely the spot," David said. "Let's hope they're here."

I bit down on my regulator, grabbed my underwater camera, and fell backward into the narrow entrance of the island lagoon. The others followed, and we descended 70 feet (20 meters) to the bottom, where we lodged ourselves against rocks and sought solid handholds. Streaming through the water, the morning sun revealed the bright reds, greens, and purples of the corals around us. A giant manta ray (*Manta birostris*) skimmed by, perhaps 10 to 12 feet (3 m) across. A green turtle (*Chelonia mydas*) gave us a sideways glance, as if curious, but too polite to stare.

Just a few minutes after finding our handholds, we felt it. Like the start of a breeze, the water began to move. Nearly imperceptible at first, the rising current gradually pulled our bubbles away at an angle as they ascended. The flow increased steadily, and a roar replaced the peaceful silence as water began to gush out of the lagoon into the ocean on the full-moon ebb tide.

Cued by this rising current, a school of perhaps five thousand green and pink Pacific longnose parrotfish (*Hipposcarus longiceps*) gathered around us and started to circle. Our bubbles were flowing sideways now as we clung to bottom rocks, and our hair and dive gear flapped and fluttered in the torrential tide. If we had lost our handholds on the rocks, we would have been swept out into the open ocean.

The foot-long parrotfish tightened their school and swam faster. This was what we had come to see: the spawning of the parrotfish on the outgoing tide. Within the group, a few fish swam faster still and shook, stimulating the entire school to spiral and bolt for the surface, joined belly to belly, releasing bursts

Rob Barrel of *Nai'a* surveys one of Kanton's many fields of fast-growing but fragile acropora corals, sometimes called table or plate corals. Fusiliers (Caesionidae) school in the water column above the reef, feeding on plankton. Smaller coral eaters, such as the chevron butterflyfish (*Chaetodon trifascialis*), live on and under plate corals, which also provide shelter or refuge for a vast variety of coral reef fishes. (Cat Holloway)

of eggs and sperm like biological fireworks. The egg and sperm clouds were so dense they dulled the sunlight that had been streaming down through the water.

For almost an hour the fish repeated this act, spiraling toward the surface every 10–15 seconds, relying on the tide to carry the fertilized eggs far out to sea, where they would be safer from predators. As I watched from the seafloor, a large shadow passed over me. A half-ton manta ray, hovering majestically and somehow unmoved by the current, was feeding serenely on the parrotfish eggs and sperm. Too soon, our nearly empty air tanks forced us back to the surface and our waiting skiff.

"Incredible! I've never seen anything like it!" said David, a coral reef biologist who has logged thousands of hours underwater studying marine life. I was also deeply moved. I have made it my goal to find Earth's last pockets of "primal" ocean, those underwater havens that have remained unspoiled for eons. In this lagoon we had found such a place.

SIDEBAR 1.1: Parrotfish Spawning · *Gregory S. Stone*

Spawning parrotfish (*Hipposcarus longiceps*). (Cat Holloway)

Parrotfish (belonging to the family Scaridae) are among many species that use a strategy called a spawning aggregation to reproduce. Large groupers, like the Nassau grouper (*Epinephelus striatus*), are also aggregate spawners. This form of reproduction seems to have evolved in fish that are normally too spread out to reproduce successfully; within an individual fish's range, mates may be few and far between. Fish form spawning aggregations on annual, seasonal, or lunar cycles, and they tend to assemble in places where strong currents can disperse the eggs. Dispersal keeps the eggs away from the large number of predators on the reef, increasing the number of fry, or young fish, that will survive to maturity.

While this form of group spawning is a successful evolutionary solution to the problem of finding a mate, it makes aggregate spawners highly vulnerable to fishing by humans. By the mid-1980s the Nassau grouper had been so depleted by fishers attracted to their spawning aggregations—when close to 100 percent of reproductive adults can be taken out of the water at one time—that it became locally extinct in many places where it had once been abundant. The Nassau grouper was one of the first fish species to be listed on the International Union for Conservation of Nature (IUCN) Red List as endangered.

As the practice of choosing your dinner live from a restaurant fish tank has become popular, the massive worldwide demand for fish has led to the targeting and decimation of many species of aggregate spawners in Southeast Asia, the western Pacific, and the eastern Indian Ocean.

To reach this remote spot, we had motor-sailed north for five days from Fiji to reach Kanton, one of eight small islands in the little-known Phoenix archipelago. Strung like jewels on an irregular necklace, the Phoenix Islands and the water surrounding them cover over 250,000 square miles (400,000 km²) of the Pacific and are part of the Micronesian nation of Kiribati (pronounced "Kee'-ree-bas").

To say that the Phoenix Islands are not easy to get to is an understatement. These islands are about 1,600 to 2,400 miles (2,963 to 4,445 km) from their nearest neighbors, with Hawaii to the north, Fiji to the south, and Samoa to the east. There are no airports, hotels, or grocery stores. The nearest airport is at Bonriki, on South Tarawa in the Gilbert Islands—946 miles (1,753 km) away. There is only one small village in the entire Phoenix archipelago, populated by a few dozen people paid by the Kiribati government to live there.

But how we reached this spot was more than a matter of planes, trains, and boats. This expedition had its seeds in a question that has long motivated adventurers: Where can we go that hasn't been explored?

Cat Holloway and Rob Barrel—explorers, filmmakers, and owners of Nai'a Cruises in Fiji—were up to this challenge. They first went to the Phoenix Islands in 1999 when their boat, *Nai'a*, was chartered by The International Group for Historic Aircraft Recovery (TIGHAR), a group of historians searching for Amelia Earhart's lost aircraft in the Phoenix archipelago. The members of the TIGHAR expedition didn't find Earhart's Electra, but as Barrel and Holloway dove these magical waters, they realized they had discovered something just as amazing: this underwater Eden was the last remaining ocean wilderness.

As the importance of their find dawned on them, Rob and Cat began to plan their return. Only a full survey could tell whether they had really found one of the planet's few unspoiled ocean oases. At the same time, their excitement was tempered by concern: if this really was a rare pristine reef, there was no time to waste. Shark finners and industrial commercial fishing vessels might find the reef and strip it of its riches in a matter of weeks. As *Nai'a* returned to Fiji, Rob and Cat were already making plans to enlist help from the other side of the world and campaign for the archipelago's protection.

At the time, I was in Fiji conducting research. When Rob got back to Fiji, he called me to find out whether I was interested in going on this first scientific expedition, and whether I could assemble a team of scientists to explore and document the marine riches of the region. To me, the question wasn't whether to go; it was how soon.

My passion for the ocean began when I was a child captivated by the ocean documentaries of Jacques Cousteau. I spent many hours with my mask and flippers on, sprawled on the living-room floor, imagining myself as Cousteau diving some distant underwater wilderness. Now, as an experienced marine biologist and veteran scuba diver, I had explored every ocean of the world via scuba and submersible. The opportunity to explore a chain of islands that had never been scientifically studied was a dream come true.

The first expedition weighed anchor in Fiji on June 30, 2000. That expedition launched the ongoing Primal Ocean Project, a quest to identify, document, and protect the planet's remaining unblemished underwater frontiers.

SIDEBAR 1.2: Looking for Amelia · *Gregory S. Stone*

Amelia Earhart. (National Air and Space Museum, Smithsonian Institution [SI 2004-11247])

As I walked the shore of Nikumaroro's lagoon, I found a wheel rim grown into the reef. At the time, I missed its potential significance, thinking it was part of the ship I had seen wrecked nearby or debris from some World War II–era occupation. I knew about the efforts of The International Group for Historic Aircraft Recovery (TIGHAR) to solve the mystery of Amelia Earhart's disappearance, but because the wheel was so easy to see, I assumed TIGHAR already knew about it. Only later did I learn that the wheel I'd spotted was a new find, and possibly significant. The wheel's size and shape seem to match that of the 1937 Electra aircraft Earhart had been flying when she and her navigator Fred Noonan vanished. TIGHAR launched another expedition in 2003 to recover it, nicknamed the "Wheel of Fortune," but the shifting sands from the pounding South Pacific Ocean had moved, buried, or crushed the find, and they could not relocate it. Details of this wheel and the search for Earhart are well told at the TIGHAR website and in the books *Amelia Earhart's Shoes* and *Finding Amelia*.

During that initial 21-day expedition, I was so astounded by the abundant, apparently undisturbed marine life that I spent the next two years organizing a return expedition with the sponsorship of the New England Aquarium, the National Geographic Society, and other supporters.

Kiribati is classified as a microstate, a designation for countries with very small populations and landmasses, 280 square miles (726 km²) in this case. But the enormous ocean area controlled by Kiribati—1,370,300 square miles (2.2 million km²)—makes it an important player on the world stage, especially as ocean resources and issues become more prominent in international affairs.

Most of the approximately 100,000 people of Kiribati don't live in the Phoenix Islands. All but a few live 800 miles (1,481 km) to the west in the Gilbert Islands or 800 miles to the east in the Line and Christmas Islands. A string of five islands, three atolls, and two submerged reefs in the central Pacific, the Phoenix Islands lie just 5 degrees south of the equator. Only one of the eight islands, Kanton, is inhabited.

Most people have never heard of Kiribati, let alone the obscure Phoenix Islands. For the early canoe explorers who originally mapped the South Pacific some 1,000–2,000 years ago, the habitat of the Phoenix Islands compared poorly with the lushness and natural resources of larger and higher Pacific islands, such as Fiji, Tahiti, Hawaii, and Samoa. Archaeological evidence reveals that there have been a few Polynesian or Micronesian settlements in the Phoenix Islands, but because of the islands' isolation from larger population centers and their limited supply of fresh water, these early settlements were only temporary.

Although the islands and the surface waters have been traversed, explored, and settled before (as detailed by Christopher Pala in chapter 2), our team was the first to scuba dive among the entire archipelago's reefs and lagoons. It was an extraordinary sensation to realize that we may have been the first humans to venture to these depths, to see the corals that have been growing here for millennia and to swim among the fish, turtles, and sharks who made this their home.

Planning the Second Expedition

After returning to the New England Aquarium, I organized a second expedition to the Phoenix Islands in July 2002. We would travel and live aboard the 120-foot steel motor-sailing vessel *Nai'a*. Along with Rob, Cat, an able Fijian crew, and National Geographic Society photographers Paul Nicklen and Joe Stancampiano, I had assembled an interdisciplinary and multinational team of eight scientists and conservationists to inventory the extraordinary marine life of these islands: Mary Jane Adams, M.D., retired anesthesiologist, medical officer; Gerard R. Allen, ichthyologist, Conservation International; Steven L. Bailey, ichthyologist, New England Aquarium; Alistair Hutt, marine mammal biologist, New Zealand Department of Conservation; Sangeeta Mangubhai, coral reef ecologist, World Wildlife Fund; Paul Nielson, representing the Kiribati government; David Obura, coral reef biologist, Coastal Oceans Research and Development in the Indian Ocean (CORDIO); and my wife, Austen Yoshinaga-Stone, bird and mammal biologist, New England Aquarium.

We set out with extremely high expectations and an ambitious science agenda. During our five-week expedition, we planned to log over a thousand scuba dives, conduct a census of bird and sea turtle nesting areas on the islands, explore the deep sea and deep reefs with special cameras and remotely operated vehicles, and survey the marine mammals in the area. If all went well, we would discover new fish and coral species and begin the work of creating the first biological map of the marine life in this remote island chain, which we believed to be the last unexplored oceanic coral archipelago in the world.

During the 2000 Phoenix Islands expedition, we deployed a deep-sea Argo camera seven times. Each time the camera descended as far as 3,000 feet (914 m). During three of the descents, my colleagues and I were thrilled when the bait pole used to attract marine animals was ripped from the rig by a six-gill shark.

Sixgill sharks live only in the deepest zone of the ocean, and they had never been previously reported from the central Pacific. First described in 1788 by French naturalist Pierre Joseph Bonnaterre, *Hexanchus griseus* remains relatively little known to shark specialists. It's a member of the Hexanchiformes, a group of sharks that has retained a number of primitive features found in fossil sharks: a single small dorsal fin, extra gill slits (most living sharks have five), a primitive digestive system, and the lack of a special membrane over the eye. The female shark retains egg capsules in her body until they hatch, and litters can include over a hundred pups.

Sixgill sharks are under increasing pressure from sport and commercial fishing and, because they reproduce slowly, can be easily overharvested. Because we know so little about sixgill sharks, they may be in more trouble than their official "low-risk" conservation status would suggest.

Sixgill sharks (*Hexanchus griseus*). (New England Aquarium)

I hoped that I might finally find in the Phoenix Islands what I had long been seeking: a coral reef environment to serve as a benchmark for restoring degraded hard coral ecosystems elsewhere. Reef ecosystems worldwide are under intense pressure: coral reefs are being degraded by destructive fishing methods and warming seas, their waters polluted by coastal runoff, and their fish depleted by intense harvesting. In the Phoenix Islands, it seemed that we had found a nearly pristine reef system, all but untouched by human hands, which might help us imagine what the rest of the world's coral reefs could be like if protected and what all of the ocean may have been like a thousand years ago before people began to deeply alter the natural world.

But mere days into our second expedition to the Phoenix Islands, it was clear that all was not well in paradise.

Vanishing Sharks

On our original Phoenix Islands expedition in 2000, we had seen sharks in profusion throughout the islands. Shark numbers have plummeted around the globe due to overfishing for their meat and fins. The abundance and diversity of sharks we had seen attested to the islands' exceptionally rich and intact ecosystem.

We had found a dazzling array of the smaller reef sharks, including blacktip (*Carcharhinus melanopterus*), whitetip (*Triaenodon obesus*), and gray reef sharks (*C. amblyrhynchos*), that are common to all pristine coral reefs. David Obura, who is based in Kenya, where there is heavy shark fishing, had marveled at the novel sight of sharks in healthy profusion. We saw sharks by the hundreds, both on our exposed dives in deep water, when the sharks would come circling out of the blue, and on our explorations of the shallow inner reef, where we found the channels jam-packed with blacktips and whitetips, feeding in a frenzy on schools of small fish.

The 2000 expedition had recorded sharks at all but one site, and at some sites we had counted over a hundred sharks in a single dive. We had seen sharks at nearly two-thirds of the sites.

As we arrived in the Phoenix Islands in 2002, we hoped and expected to see the same high numbers of sharks on the reefs. Diving into the water at Nikumaroro, our first island stop, we again were greeted by a profusion of sharks after our two-year absence. But at our next stop, Manra, we were surprised to find only a few sharks in the water.

"Maybe they are on the other side of the island," I said. We motored in the skiff to the other side of the island, but still found barely any sharks. We next stopped at Rawaki and again saw very few sharks. By the time we reached Kanton and were able to talk to people, we were seriously alarmed: Where had all the sharks gone? What catastrophe could explain their plummeting numbers? At Kanton, again we found the sharks almost completely absent, and now we discovered the reason why.

In 2001 a boat en route from American Samoa stopped in the islands to catch sharks for their most valuable product—their fins. Shark fins are used as an ingredient to thicken a soup traditionally served at weddings and other cer-

A diver watches the maiden descent of a remotely operated camera being driven from the deck of *Nai'a*. The camera was used to film the Phoenix Islands' deep reefs, which lie beyond the safe limit of scuba divers. (Cat Holloway)

Whitetip reef sharks are abundant throughout the shallow waters around the Phoenix Islands. They are usually shy of divers but can become very aggressive, fast, and competitive when hunting or feeding. (Cat Holloway)

emonies throughout China and Southeast Asia, and the surging demand has pushed shark-fin fishing to the farthest corners of the globe. The fishermen keep only the fins, which can be dried, as shark meat rots faster and they cannot get it to distant markets. The dried fins are sold for up to €500 or US$630 per kilo, many times the price of shark meat. Stripped of their valuable fins, the maimed sharks are thrown back into the water to die.

The shark-finning boat that visited the Phoenix Islands stopped at Kanton, Rawaki, Manra, and briefly at Orona to catch sharks using longlines. Longlining is a commercial fishing practice in which boats spool out up to 80 miles of fishing line baited with up to forty thousand hooks to reel in fish by the thousands. The shark-finning boat stayed in the islands for about three months. Luckily for the remaining sharks in the rest of the island group, engine trouble eventually forced the shark finners to head back to Samoa. But the damage was considerable: in a few short months, this single boat removed almost the entire

SIDEBAR 1.4: The High Price of Shark Fin Soup

Shark fins for sale on the sidewalk in Hong Kong. Shark Savers estimates that 50 percent of the world trade in shark fins passes through Hong Kong. (Paul Hilton Photography)

Shark fin soup is a status symbol throughout much of Asia. For generations, only the richest of the rich could afford this gelatinous delicacy, but as Asia prospered over the last fifteen years, its growing middle class acquired an appetite for all the trappings of success—especially shark fin soup. Being able to order it, at $85 to $100 a bowl, and serve it to guests at a wedding banquet was a sure way to signal that you'd achieved the good life.

But it's not good for the sharks. As many as 100 million sharks are being taken from the ocean every year, and sharks reproduce too slowly to replace these numbers. Most of those millions of sharks are stripped of their fins at sea and thrown back into the water. Unable to swim and get fresh oxygen through their gills, they sink to the bottom of the sea and drown. Divers have reported seeing vast graveyards of finned sharks.

Efforts to end this unsustainable and wasteful practice have run into stiff opposition from countries such as Japan, which successfully blocked attempts in 2010 to extend international legal protection to such endangered shark species as the blue shark, thresher shark, and hammerhead. Activists in Hong Kong and the United States have succeeded in getting high-profile venues such as Disneyland in Hong Kong and Las Vegas casinos to take shark fin soup off their menus, but it may already be too late for some shark species, whose populations are crashing worldwide. Ominously, even as Hawaii passed landmark legislation in 2010 banning the import and possession of shark fins, reports were surfacing that operators of shark-finning boats, running out of sharks to fin, were turning their attention to the sharks' relatives, skates and rays.

adult population of sharks from half of the Phoenix Islands, including the three largest islands.

Commercial shark finning also affects the local human population. While the Samoan shark-finning boat was in the islands, I-Kiribati islanders were engaged in artisanal shark fishing based in Orona, both for the local consumption of meat and for the export of fins via small shipping businesses on the main island, Tarawa. The islanders reported that they had abundant catches at the beginning of their stay on Orona, but after a few months, their catches sharply declined.

By the time we visited Orona in 2002, there were few sharks left, and local income from shark fishing was near zero. Manra, Rawaki, Kanton, and Orona had drastically reduced shark populations. Enderbury, Nikumaroro, and Birnie still had healthy numbers of sharks, similar to those we'd recorded in 2000.

This one example of the incredible overefficiency of commercial harvesting of valuable products, in which a resource is extracted to the point of zero return, convinced us of the urgent need to protect the Phoenix Islands. This remarkable archipelago required quick and effective management to preserve its resources for the benefit of local people and ensure that those resources are never decimated by overuse. The race was on, and time was running out.

Expedition Diary: Swimming with Sharks, Nikumaroro, 2000

GREGORY S. STONE

I first came to Nikumaroro in 2000 to survey the sharks that gathered in the shallow lagoon in great numbers. Our plan was to survey sharks in the upper reef and to seek new species of other fish in the deep reef zone.

Exposed to waves on all sides, Nikumaroro offered no safe harbor for a boat of *Nai'a*'s size. We left her anchored safely off the island's western point. We would have to approach this unwelcoming island by skiff. Carrying my scuba gear and cameras, I made my way down *Nai'a*'s side deck to the dive skiffs tied off her stern. Gerry Allen, Steve Bailey (known to everyone as Bailey), and my wife, Austen, joined me in a skiff.

Our skiff sped toward the island, bouncing on passing swells like a skipping stone, and *Nai'a* disappeared from view around the point. Shadowy outlines of sharks darted beneath us in the clear water. As the skiff came to rest in the shallows 200 yards from Nikumaroro's south side, we could see the narrow lagoon entrance and palm trees jutting up through the dense undergrowth.

"OK, let's go," I shouted, and we rolled backward into the water as a group for safety.

Divers are most vulnerable to sharks at the surface and in mid-water. Many sharks attack prey from below, and I wanted to get to the bottom fast, where we would have one less direction to worry about. Austen and I tucked in among the coral heads at 60 feet (18 m), then watched Gerry and Bailey continue over the reef edge into deeper water, where they would search for reef fishes.

The water around Austen and me was filled with gray, whitetip, and blacktip reef sharks. They appeared to be hunting for food amid a school of nearly

Convict surgeonfish (*Acanthurus triostegus*), also called convict tangs or manini, school by the hundreds or thousands in shallow water where currents are strong and algae thrive on coral and rock. Agile swimmers, convict surgeonfish graze en masse on algae. Without these herbivores, coral reefs would soon be smothered by algae and die. (Cat Holloway)

two thousand yellow convict surgeonfish (*Acanthurus triostegus*) that were grazing on algae along the bottom and several hundred bigeye jacks (*Caranx sexfasciatus*) that were passing above us. Sharks generally don't attack divers without provocation, but their shape and manner can still have an unnerving effect and trigger a primeval fear, stoked by images of sharks in popular culture and media accounts of shark attacks. Even an experienced diver and scientist can feel a frisson of fear when swimming with sharks.

But we hadn't come entirely unprepared. As we moved down the reef, Austen was carrying a blunt, 2-foot plastic "shark stick" to hold off curious or aggressive animals. "False security is better than no security," she had told me back on the boat. I planned to use my underwater video housing, the size of a car battery, as a shark deterrent, if needed.

Turning back toward Austen, I saw silhouettes of sharks behind, above, and all around her. As I looked ahead, a 6-foot gray reef shark shot at me like a torpedo. I stiffened, kicked back, and thrust my camera housing toward it. The

Nai'a captain Rob Barrel goes nose to nose with a blacktip reef shark (*Carcharhinus melanopterus*). (Jim Stringer)

shark was also surprised, its focus apparently on something else, just as mine had been. It veered off and darted away like the snap of a whip, passing in a blur only 8 inches (20 cm) away from me. A shark can get confused at moments like this and bite a diver it otherwise would have left alone.

We completed our survey without further incidents. "I've never seen so many sharks!" Austen exulted as she pulled herself back into the skiff, clearly glad to be out of the water. "They're so graceful."

I was experiencing the same mixture of exhilaration and relief . . . glad to be back in the skiff, amazed and grateful to have seen over a hundred sharks on the reef.

2 HISTORY THROUGH A WATERY LENS

CHRISTOPHER PALA

Since the arrival of the first explorers from Polynesia and Micronesia in the tenth century, the availability of water has had a profound impact on the ability of humans to live on the Phoenix Islands.

Rain is a fickle visitor to the Phoenix Islands, which first rose from the ocean some 80 million years ago. When precipitation does fall, the columns of warm air that rise from the shallow lagoons split the low rain clouds as they pass above the island, so that often it rains right next to the land, but not on it. The amount of rain that falls *on* the islands varies enormously. On Kanton, for instance, rainfall totals have ranged from 112 inches (285 cm) in 1941 to 8 inches (20 cm) in 1954. Generally, the northern Phoenix Islands, closest to the equator, are more arid.

But even if a good drenching rain falls, how could you (if you were an islander) capture rainwater in a sieve—an atoll made of porous coral, sandstone, and sand? You couldn't, but nature does it for you. Here's how it works. If you pour a glass of fresh water into the sea, it will instantly mix with the salt water and become undrinkable. But the upper part of an atoll is made up of dead, porous coral, like a hard sponge. Most of it lies below sea level, but in places it protrudes above the high-tide mark a few yards. Now pour your glass of water on the exposed coral. Gravity will pull it down through the coral's myriad holes until it reaches the salt water. But absent any movement of that water, the fresh water will float above the salt water because, lacking salt and other minerals, it is slightly less dense and therefore lighter.

Looking at the atoll from above, we can see that below the emergent part, imprisoned in the tiny holes of the coral, lies what is called a lens of fresh water. The lens can be up to 50 feet (15 m) thick and follows the contours of the island—right up to the edge. Trees that grow on the beach get fresh water from their roots, even as their trunks are splashed with seawater at high tide.

Every drop of rain that falls on the island goes into that freshwater lens. As long as no one takes the fresh water out of the lens, it doesn't matter how much it rains: the fresh water, a few feet below the island's dry surface, won't evapo-

The *Hōkūle'a*, a reproduction Polynesian voyaging canoe build by the Polynesian Voyaging Society, under sail off Diamond Head in Hawaii in 1976. Voyagers in similar vessels may have reached the Phoenix Islands thousands of years ago, but left no permanent settlements. (Sam Low Photography)

rate. It will just sit there until a big storm washes waves right over the island, pouring salt water into the freshwater lens, mixing the two. Eventually, the fresh water will rise once again to the top, restoring the balance.

The First Arrivals

Imagine yourself a Polynesian voyager traveling across huge expanses of ocean on a sailing catamaran, with no instrument or chart other than ancestral knowledge, passed on orally, that tells you how to get to where you want to go by reading the patterns of stars, birds, and ocean swells. You arrive at an atoll, and the first thing you do is dig a shallow well. The water will be fresh and pure. Hurricanes are unknown in the Phoenix Islands, and storms are rare. Not a bad spot. You settle the island, and everyone begins drinking from the lens and using its water to wash and cook. As you use up the fresh water sitting atop the salt water, the salt water rises, and soon there is no more fresh water. You are drinking seawater.

That's where the rain comes in: it's what replenishes the fresh water. The more it rains, the more fresh water you can take out of the lens. But if you and your people use up the fresh water faster than the rain replaces it, or show up during a multiyear drought, you simply can't live on this island. And so you die, or you leave.

Archaeologists have found ample evidence that voyaging Polynesians and Micronesians settled on the Phoenix Islands at various times from the tenth to the sixteenth centuries, but it seems they never stayed for more than a few years.

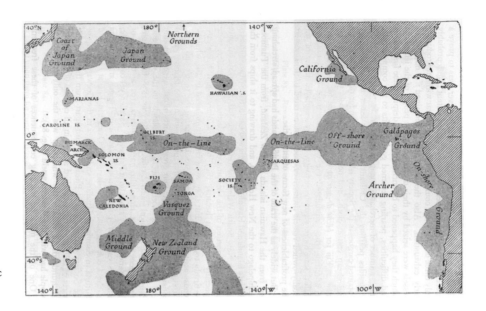

A map showing the Pacific whaling grounds of the nineteenth century. The first Europeans to sight the Phoenix Islands may have been whalers fishing "on the Line"—following migrating sperm whales along the equator in the Pacific. (The National Archives, Kew)

Sydney (now Manra), though, appears to have supported a population as high as one hundred. Those people stayed long enough to build fishponds, houses with foundations, pits to grow vegetables, and several *maraes*, Polynesian paved plazas used for ceremonies or meetings. On Hull (now Orona), they left a stone pyramid containing turtle and bird bones. On other islands, they left nothing but a few of the rats they brought for food, which multiplied and exterminated some of the bird colonies.

In *Sailing Routes of Old Polynesia*, Anne di Piazza and Erik Pearthree write that these earliest settlers left in periods of drought, but other theories suggest that they may have lived mostly on eels, squid, and shellfish along with seabirds and eggs. After depleting those resources, they simply moved on to other islands.

Just when Westerners discovered the Phoenix Islands is unclear because those who did were not professional discoverers. We know now from their journals that the Frenchman Jean-François de Galaup, Count of La Pérouse, passed 10 miles from Sydney (Manra) in 1787 and that the Russian explorer Otto von Kotzebue sailed within 55 miles of Enderbury and then Kanton in 1816. But the captains who first laid eyes on these islands were whalers who were fishing "the Line," as the nutrient-rich equatorial region was called, probably between 1800 and 1820, and they didn't bother describing the uncharted islands they happened to sail by. Indeed, they were secretive about the location of their fishing grounds. The Phoenix Islands were not among their favorites because, except for Kanton (then spelled Canton), they offered no safe anchorages. When the islands' sizes and positions were finally noted, it was in the ship's log of an expedition lasting from 1838 to 1844 and led by Lieutenant Charles Wilkes, a man whose faith in the efficacy of the cat-o'-nine-tails is believed to have made him the model for Captain Ahab in Melville's *Moby Dick*. The "United States Exploring Expedition to the Wide Pacific Ocean," as it was officially called, was the belated American answer to the European explorers. It was also the last circumnavigation to be made entirely by sail.

Meanwhile, trends were taking shape on the other side of the planet that would bring far more change to the Phoenix Islands than either the Polynesians or the whalers had caused.

The Great Guano Rush

At the start of the nineteenth century, as their crop yields dropped dramatically, American farmers were discovering the limits of unfertilized soil. Accumulated seabird droppings known as *huano* (later called guano) had been used by Peruvian farmers for centuries to turn some of the world's most arid soils into a reliable provider of potatoes and vegetables, the economic bedrock of the Inca empire. As interest in Peruvian guano grew in Europe and in the United States, its supply, tightly controlled by the Peruvian government, did not, and prices remained high, right where the Peruvians wanted them.

This situation set off a global guano rush to find alternative supplies. Big seabird colonies, where guano deposits could be dozens of feet deep, had been decimated around the world, not so much by people but by the rats and cats that came with them. So the focus fell on uninhabited, remote islands, most of them in the Pacific. As American farmers and the politicians who represented them clamored for a secure, reliable, and cheap source of guano, the Phoenix Islands and other islands in the region became targets for the guano trade.

SIDEBAR 2.1: The Great Guano Rush

Guano was so highly prized by the ancient Incas that disturbing nesting birds was an offense punishable by death. During the nineteenth century, Peru grew rich on its vast guano stores. By the 1850s, guano was so important that the subject even made it into the 1850 State of the Union address of President Millard Fillmore: "Peruvian guano has become so desirable an article to the agricultural interest of the United States that it is the duty of the Government to employ all the means properly in its power for the purpose of causing that article to be imported into the country at a reasonable price."

Why was there so much concern about bird droppings? During the Industrial Revolution, industrial powerhouses like England and the United States became eager to make their agriculture more efficient to reduce their reliance on imported wheat. Guano was an ideal fertilizer; it could also be refined into saltpeter, a component of gunpowder. The problem was that Peru had a virtual lock on the world's guano stores—and the United States was eager to find and exploit its own supply. In 1856 the U.S. Congress passed the Guano Island Act, allowing the United States to take "peaceable possession" of uninhabited islands with guano deposits. The Great Guano Rush came to Enderbury in 1870, when the Phoenix Guano Company sent a superintendent and some sixty Hawaiian laborers there. Over the next seven years, more than 70,000 tons (6.4 million kg) of guano were shipped from Enderbury alone. The United States finally relinquished its claim on the Phoenix Islands in 1979, when, with the Gilbert and Line Islands, they became the Republic of Kiribati.

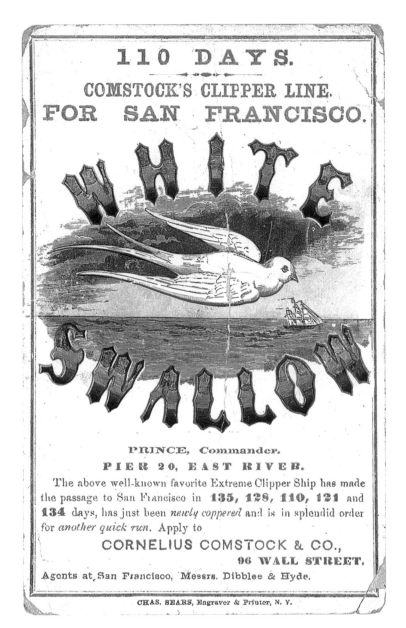

In 1859 the clipper ship *White Swallow* brought 1,200 tons of guano from McKean Island to New London, Connecticut. (Courtesy of the collection of Bruce Roberts)

To encourage this industry, the U.S. Congress passed the 1856 Guano Islands Act, allowing any American to take over unclaimed and uninhabited islands containing guano and claim them as U.S. territory, as long as they were continuously occupied and the guano was sold in the United States. The government, in turn, promised to defend those entrepreneurs against foreigners. The United States kept the option of divesting itself of the islands after the guano had been mined out.

The law, which remains on the books, triggered a rash of claims, sometimes for islands that didn't exist, often multiple claims for the same island. Companies were often more intent on claiming islands than on occupying them and mining the guano. As the U.S. State Department would recognize one company as the owner of an island, another would quietly slip in and mine and export its guano undisturbed. And while the bureaucrats were firm in dealing with the claims of other countries, appeals by one company asking the government to throw another off some distant speck of land went mostly unanswered.

In the end the Great Guano Rush brought about 400,000 tons (360 million kg) of fertilizer to the United States, nearly a quarter of it from Enderbury. Eventually, the fertilizer magnates found great domestic deposits of mineral phosphates in Florida. By enriching this native American resource with additives—ranging from slaughterhouse offal to oil by-products—to provide the required nitrates, they created a new, non-guano fertilizer that made bird droppings obsolete.

Trade was not particularly profitable for the big guano companies. Life on the islands was miserable for the workers, mostly native Hawaiians. Some sued their employers in Honolulu courts for breach of contract and back pay, but the courts sided with the companies. Water was just as scarce as it had been for the earlier Polynesians and had to be distilled from seawater with inefficient stills or imported at great cost.

By 1878 the Phoenix Islands had returned to their usual state, visited by the occasional ship and saddled with the occasional shipwreck.

Outgrowing the Gilbert Islands

The next settlers came a half century later from the Gilbert Islands, some 800 statute miles (1,481 km) to the east. The total land area (think freshwater lens) of the Gilberts, at 106 square miles (171 km²), is ten times that of the Phoenix Islands.

The population of the Gilberts initially fell after Westerners introduced diseases to which islanders had no immunity. By the 1920s, with the advent of modern medicine and the end of island warfare, the population had recovered and far exceeded its precolonial size. This new era of health and peace brought with it intense competition for land. Harry Maude, the islands' administrator, acknowledged that this "land hunger" was a problem "we ourselves had largely created." He resolved to find new islands in the region where the Gilbertese could resettle.

After a long search of the British Empire in the Pacific, Maude decided that the best fit for the landless Gilbertese was the nearby Phoenix Islands. After all, he reasoned, the Gilbertese way of life centered around the coral atoll, so the settlers would be ill suited to life on high, forested volcanic islands. In addition to reef seafood, gathered by women on foot at low tide, and fish, caught by men in canoes, they lived mostly off the coconut and the pandanus trees, having long ago exhausted as a food source the seabird colonies they had found when they arrived.

After exploratory trips to all eight islands, Maude and his team determined that Hull, Sydney, and Gardner (now Nikumaroro) were most suitable. In January 1939 they sailed over with 61 poor, land-hungry immigrants, leaving 10 on Gardner, 10 on Hull, and 41 on Sydney. The program turned out to be the last act of territorial expansion of the British Empire.

By September 1940 a total of 729 colonists had been taken to the Phoenix

Lack of fresh water has always limited the human presence in the Phoenix Islands. In the 1930s British colonial administrators attempted to settle the islands, but by the early 1960s the scheme was abandoned and the settlers relocated. In 2001 the Kiribati government once more tried to establish a settlement on Orona. These villagers were photographed on Orona in 2002. (Cat Holloway)

Islands, of whom only 7 had been repatriated. Sydney and Hull, Maude reported, were "normal self-contained island communities," selling their dried coconut, or copra, to cooperatives set up by the British government. Gardner had become the headquarters of the new Phoenix Islands District. Phoenix (now Rawaki), Birnie, and McKean, home to millions of nesting seabirds, were declared tributary islands to the colonized ones, to be raided for food when needed.

Colonization resumed after World War II, and the population peaked at over 1,300 in the mid-1950s. But drought came and discontent rose. By the early 1960s all the Phoenix Islands had been evacuated by the British authorities. Most of the settlers moved to the Solomon Islands, where they remain today.

The Aviation Era

In 1928 the Australian Charles Kingsford Smith staged the first transpacific flight, passing directly over the Phoenix Islands. Their convenient location halfway between Honolulu and New Zealand made the islands suddenly attractive as refueling depots.

In 1935 Pan American Airways selected Kanton as a stop for its luxurious Clipper seaplanes, which took four days to fly from San Francisco to Auckland. Lack of water was not a deterrent: Pan Am could afford to desalinate seawater at great expense. A hotel was built, the lagoon was dredged and deepened to become a watery runway, and in July 1940 the first Clipper splashed down. But the attack on Pearl Harbor in December 1941 abruptly ended civilian flights. Kanton was taken over by the U.S. military, which built a runway and turned the island into a hub of Pacific aviation, with up to 1,100 people present on the island at one time.

Between July 1940 and December 1941, Pam American Airways flew Clipper Ship Boeing 314s to New Zealand, stopping over at Kanton (then Canton). Flights were briefly resumed after the war before competition from long-range jets prompted Pan Am to ground its "flying boats." (Courtesy the Radio Heritage Foundation)

After the war Pan Am retired its fleet of seaplanes and returned to Kanton with a fleet of wheeled, propeller-driven airliners. Along with other airlines, it used the island as a stopover for flights between Hawaii and Australia and New Zealand. Kanton hosted up to 12,000 passengers a year. A fish export business flourished, sending up to 8 tons a month to Hawaii. But in 1959 the advent of the long-range Boeing 707 made the Kanton stopover obsolete, and the hotel and the rest of the Pan Am operation closed down.

The Space Age and the Cold War

Kanton promptly reinvented itself as a NASA tracking station. From 1959 to 1967, a staff of five hundred provided support for the Mercury space program, which sent the first Americans into space. Designated Site 11 by NASA, the station on Kanton monitored spacecraft during the crucial reentry phase, communicating with Mission Control by teletype and voice-loop networks. NASA pulled out in 1967. Four years later, the Americans returned, this time in the form of a U.S. Air Force outpost of the Space and Missiles Systems Center (SMC).

The U.S. military was firing Minuteman missiles from California at test targets in the remote Pacific, and Kanton offered a convenient monitoring station.

By 1979 the United States had pulled out altogether, and Kiribati was born as an independent nation encompassing the overpopulated Gilbert Islands, the unpopulated Phoenix Islands, and the Line Islands.

Ever since, the Phoenix Islands have been left largely to their own devices, with a handful of functionaries and their families serving in Kanton to maintain its runway and ensure Kiribati's claim to the archipelago.

3 PARADE OF THE BUMPHEADS ORONA

GERALD R. ALLEN AND STEVEN L. BAILEY

As the two fish biologists with the 2002 Phoenix Islands expedition, we were the onboard experts with encyclopedic ichthyological knowledge. Seated around the table in *Nai'a*'s cabin, we engaged in endless "bio-babble," swapping the scientific names of various species with abandon, while the rest of our shipmates looked on, amused and mystified. Normally, we were each compelled to keep such fishy thoughts strictly to ourselves, and it was wonderful to have a like-minded soul to share them with.

On arriving at Nikumaroro Island, we quickly settled into a routine of three to four dives each day. Our choice of dive site was influenced by several factors. For the best collecting results, we needed to avoid spots with strong tidal currents. We also tried to sample a range of habitats, from sheltered lagoons and passes or channels to outer reefs. Our dive schedule was dictated by the need to space out the dives to avoid decompression. Normally we would have a two-hour interval between dives, which we spent processing fish we'd collected and entering data from the previous dive. Our primary goal was to conduct a comprehensive survey of the reef fishes, from the shallows down to about 165 feet (about 50 m), the lower limit of safe scuba diving.

We were excited about the fish prospects and hoped to substantially boost the total fish count for the islands, which previously stood at nearly 300 species. In 2000 Bailey had collected and observed a wealth of fishes, including 107 new records for the island group. In combination with 184 species collected by Dr. Leonard Schultz aboard the USS *Bushnell* in 1939, this effort formed a solid foundation of knowledge on which we hoped to build. There was also the tantalizing prospect of being the first to describe a previously unknown species.

Cataloging fish is similar to bird-watching, only with an air tank and regulator. There are many more species, but far less time to record them due to the constraints of diving. We would rapidly descend to about 100–165 feet (about 30–50 m), then slowly rise to the shallows, recording fish species as we went. We spent most of our observation time at depths between 6 and 50 feet (2 and 15 m), as this biologically fertile "sunlight" zone harbors the highest number of

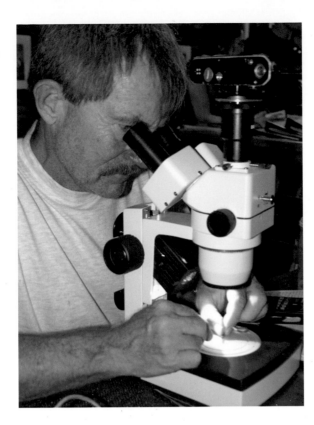

Gerry Allen identifying fishes at
the microscope on board *Nai'a*.
(Cat Holloway)

species. Using an underwater clipboard and data sheet, we would record the species we observed. Our accuracy during this activity relied entirely on our collective experience, which consisted of thousands of hours underwater between us. As with bird identification, the body shape and size, color, and general behavior of fishes provide important cues for correct identification. After each dive we would compare notes and discuss any species we had difficulty identifying.

The diving was absolutely superb—every bit as good as we had hoped it would be—often with a hundred feet of visibility and a wide array of colorful reef fishes at every site. Most were familiar species that Gerry knew from other locations in the Indian and Pacific Oceans: convict surgeonfish, Moorish idols, and yellow longnose butterflyfish range widely across this vast region.

It was a special pleasure to see a pair of orangefin anemonefish (*Amphiprion chrysopterus*). Gerry had studied their life history as part of his Ph.D. dissertation. These fish are common throughout the Phoenix Islands, and during a dive at Birnie Island, we encountered a pair that were spawning. Luckily, we had our cameras in hand to record the action. The female deposited a patch of tiny orange eggs that were summarily fertilized by her partner. As a result of his past research, Gerry knew that the parents would busily care for the nest over the next week, until the eggs hatched and the larvae were carried away by the currents.

What impressed us most was the overwhelming numbers of individuals of many species. Fishes that we rarely saw in the waters off Indonesia and the Philippines were visible here in spectacular abundance. On most dives, we were

IDENTIFYING REEF FISHES

Reef fish species can be confusing, with scientific names that often defy pronunciation and common names that may sound alike. Fish biologists rely on close observation of a number of features for identification: the types and placement of spots and stripes, the numbers of dorsal spines, and the position and shape of fins and tail. In many cases an experienced ichthyologist can identify a fish on sight. In other cases even professionals have to resort to a microscope or even a tissue sample to make a final identification.

Left to right, three common reef fishes: a convict surgeonfish (*Acanthurus triostegus*); a bigeye soldierfish (*Myripristis vittata*), and a bigeye trevally (*Caranx ignobilis*). Even a novice fishspotter can see the clear differences in body outline, the fishes' lateral lines, and the presence and location of dorsal spines.

Here are two superficially similar species with confusingly similar names: the bumphead parrotfish (*Bolbometopon muricatum*) and the humphead wrasse (*Cheilinus undulatus*). Their identifying features have been labeled.

Bolbometopon muricatum
Size: Common length 70 cm, max 130 cm
Note that the "bump" is more vertical that the wrasse's "hump." Exposed tooth plates form the "beak" that give the parrotfish its name.

Cheilinus undulatus
Size: Common length 60 cm, max 220 cm
Note the two black lines behind the eye.
Elongate dark spots on scales tend to form bars.

Data: www.fishbase.org. *Images*: *A. triostegus*, Joanna Woerner/IAN Image Library; *M. vittata*, *C. ignobilis*, *B. muricatum*: Tracey Saxby/IAN Image Library; *C. undulatus*, Dieter Tracy/IAN Image Library. IAN Image and Video Library, ian.umces.edu/imagelibrary/.

greeted by swirling masses of black jacks (*Caranx lugubris*) and bigeye jacks. There were also huge numbers of surgeonfishes (Acanthuridae), particularly at Nikumaroro.

The majority of reef fishes have open-ocean or pelagic egg and larval stages, which are widely dispersed by winds, waves, and currents. The yellow longnose butterflyfish (*Forcipiger flavissimus*), for example, which was common in the Phoenix Islands, ranges all the way from Panama to East Africa.

Near the end of an otherwise routine dive at Orona, we spotted a procession

Lettuce corals were one of the most common coral formations on the reefs, providing habitat for many types of fish and invertebrates. (Brian Skerry)

of green bumphead parrotfish (*Bolbometopon muricatum*). These are not just ordinary parrotfish, but one of the reef's truly gigantic creatures, reaching a length of over 4 feet (over 1 m) and tipping the scales at well over 100 pounds (about 45 kg). Usually, people see them in relatively small groups of fifteen or fewer fish. But at Orona we watched in amazement as a school of several hundred fish paraded by over several minutes, calling to mind a vast herd of buffalo.

The Napoleon or humphead wrasse (*Cheilinus undulatus*) was another giant we recorded at most dive sites. Highly prized in the live fish trade, this fish supplies restaurants in Hong Kong and other Southeast Asian centers. It can reach as much as 7.5 feet (2.2 m) in length and weigh 420 pounds (190 kg). An individual fish can fetch thousands of dollars in the large wholesale fish markets of Hong Kong and mainland China. Consequently, the species is rapidly disappearing over much of its range, particularly in Southeast Asia. Surveys by Conservation International (CI) in Indonesia since 2001 and in the Philippines between 1998 and 2009 recorded a total of only 20 sightings during 150 dives. In marked contrast, we saw between 20 and 25 large Napoleons on nearly every dive in the Phoenix Islands. There was clearly an urgent need for conservation action, as the population could be quickly devastated if foreign or local ventures started fishing operations here.

Like other long-lived species, the Napoleon wrasse reproduces slowly, which makes it hard for the species to recover from the pressures it faces: intensive fishing for the live fish trade, recreational spearfishing, destructive fishing with cyanide and dynamite, and the removal of young wrasse for aquaculture, in addition to the larger pressures of habitat loss and climate change. The Napoleon wrasse is listed as an endangered species on the International Union for Conservation of Nature (IUCN) Red List, but is still actively pursued to supply the Asian

PHOTO ESSAY: Why Are Reef Fish So Colorful?

Scientists have long realized that color plays an important role in the animal world. Color is used as an advertisement to mates: potential fathers broadcast their desirability to females using gaudy colors. In birds, pigments called carotenoids—the reds, yellows, and oranges—appear to play a particularly complex role. These powerful antioxidants seem to accurately reflect the male's sperm quality and promote the health of developing chicks in the egg, as well as playing a role in bird vision. At the same time, bright pigments are used widely in the animal kingdom to warn predators of toxicity. Nontoxic animals even mimic the colors and patterns of unpalatable species to avoid being eaten.

Coral reef scientists looking at the psychedelic color scheme of tropical reefs and their residents initially assumed that these same two phenomena explained color in tropical fish as well. But recent research from around the world suggests there may be more to the story.

Pseudanthias. (Cat Holloway)

Tomato grouper (*Cephalopholis sonnerati*). (Gerald Allen)

Orangefin anemonefish (*Amphiprion chrysopterus*). (Jim Stringer)

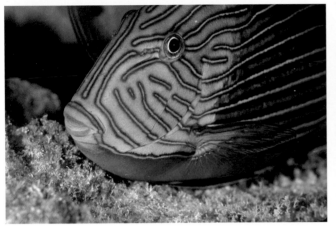

Lined surgeonfish (*Acanthurus lineatus*). (Paul Nicklen)

Arch-eye hawkfish (*Paracirrhites arcatus*). (Mary Jane Adams)

In 2004 researcher Les Kaufman of the New England Aquarium and Tim Laman dove the reefs off Fiji and Indonesia to try to determine what was going on with fish color. Kaufman had already realized that color was important in the clear, sunlit waters of the reef. In the turbid waters of the deep ocean, fish are much less colorful, and they rely on other channels of communication, such as taste, touch, and smell. The clear waters of the reef clearly opened up a new visual channel for sending messages. But what messages, exactly, were the fish sending?

Advertising to mates and warning predators of toxicity are still clearly important underwater, but reef creatures were using color in many more ways, taking advantage of the ultraviolet light channel we don't see with our limited primate vision. Tuning in to this UV channel with special equipment allowed Kaufman and other researchers to see what the fish saw. Or rather, what they didn't see.

Seen in UV, the color patterns on some reef fish turn out to be highly effective not at making fish stand out, but at helping them blend in. Much in the way the stripes on a zebra break up its form, stripes and polka dots on reef fish break up their outlines and help them disappear. The ability to see in the UV light spectrum might also help fish that feed on plankton find their tiny prey in the water column.

Parade of the bumpheads. (David Obura)

One of my favorite fishes in the Phoenix Islands is the green bumphead par- rotfish (*Bolbometopon muricatum*). These 4-foot (more than 1 m) giants roam the reef in huge schools, grazing on the corals like buffalo. I like this fish be- cause it's colorful, it's huge, and its physiology is in a class by itself. The sound of its huge jaws crunching coral is, to use my son's jargon, wicked awesome. Gerry Allen and I saw large schools of green bumpheads in 2000 and 2002—a welcome sight, because these tasty fish are under a great deal of pressure from overfishing elsewhere in their range.

I also like green bumpheads because of their fascinating relationship with the corals on which they feed. It's easy to see what the bumphead gets out of the corals: a plentiful food source. What's less obvious is what the corals get out of the relationship. After all, a single mature *B. muricatum* can crunch its way through 5 tons of coral per year, of which up to 2 tons typically includes living coral tissue. Bumpheads excavate the hard carbonate skeleton with their massive front teeth and grind it up with additional teeth farther back in the throat, called pharyngeal mills. They also use the large hump on the head to ram the coral, breaking it into smaller pieces. This would seem to be a lose- lose proposition for the coral. But while individual live corals lose in the short term, in the end, reefs may win.

Reefs in places like the Phoenix Islands need grazers like the bumpheads to keep them healthy. Counterintuitive as it might appear, the constant trimming makes for a healthier coral garden. It seems that by breaking down coral skel- eton (a process called bioerosion) and redepositing it on the reef as a kind of cement, the bumpheads make the reef stronger and better able to withstand the high-energy waves of these remote atolls.

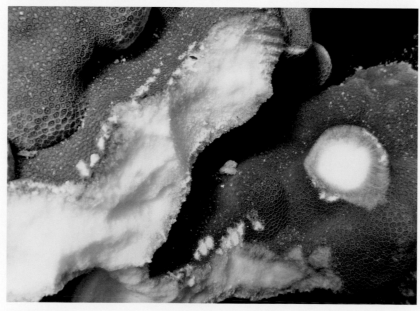

(Randi Rotjan)

Many of the reef fish we photographed may never have seen a diver before, let alone a camera. (Wendy McIlroy)

market. Moves by Australia, China, and other nations to protect the species are encouraging, but it has yet to earn endangered species protection in the United States. For now, the Napoleon wrasse is a U.S. National Marine Fisheries Service Species of Concern.

Gerry was especially interested in the composition of the damselfish community. These small, colorful reef dwellers had been a focal point of his research since the early 1970s. The family Pomacentridae contains more than 350 species, which are among the most dominant daytime fishes on coral reefs worldwide, both in number of individuals and number of species. In the Phoenix Islands we recorded 36 damselfish species. At the outset, Gerry was optimistic about finding a new species of damselfish here, since the deep unexplored reefs seemed to be an ideal location for an as yet unknown species to be hiding. But after three full weeks of intense searching, we still had not come up with our prize.

The search for the elusive new damsel reached desperation levels by the time we arrived back at Nikumaroro. Only two days remained before our scheduled return to Fiji. We decided that our best chance was to make a deep dive on the wave-battered seaward slope, so we decided to do a recon, leaving our collecting

Allen and Bailey collecting samples in the lagoon of Orona. (Cat Holloway)

equipment behind. There was still enough time for a collecting dive if we struck gold. So we set off with just our cameras. At first, aside from some very curious gray reef sharks, our recon dive was uneventful.

We slowly worked our way down the steep slope to a depth of 150 feet (about 47 m). Suddenly, there it was, right in front of Gerry—a small, inconspicuous fish, darting among the rubble. Gerry nearly spat out his regulator, gesticulating wildly and shouting to attract Bailey's attention. After a few quick photos, our dive computer alarms announced it was time to head back to the surface.

We got little sleep that night as we constantly reviewed and revised the tactics for tomorrow's critical collecting mission. The pressure was definitely on. This was going to be the last dive of the trip, and although damsels are normally easy to catch, Gerry would have precious little bottom time, about eight minutes maximum at a depth of 165 feet (50 m). We planned to use squeeze bottles filled with a fast-acting anesthetic, quinaldine sulfate, which is very effective once the fish is holed up in a blind crevice. Once anesthetized, the fish is easily captured with a small hand net. The next morning at breakfast, we went over our plan one last time. We were ready for the hunt.

Although we used GPS coordinates to locate the site, we had difficulty finding our bearings as we descended the slope. It slowly dawned on us that something was wrong. Yesterday the new damsel was common below 150 feet, but now we didn't see a single one. Eventually we crossed over into an adjacent ravine and were greatly relieved to find several. Gerry chased one into a crevice and squirted a liberal dose of anesthetic. The fish emerged in a drugged stupor and swam straight over to Bailey, taking refuge in his buoyancy vest. Bailey hadn't noticed the fish's destination and thought Gerry had gone berserk when Gerry grabbed him and nearly tore the vest off. The fish was nowhere to be

SIDEBAR 3.2: Tools of the Trade

(Jim Stringer)

On an eleven-day expedition to the Phoenix Islands in 2009, researchers made a total of thirty-nine dives to take measurements, check the condition and health of the reefs, and census marine organisms. In order to take advantage of their seventy minutes underwater, our team needed the right equipment. In addition to basic scuba gear and wetsuits, here's a list what they brought along to get the job done.

30-meter transect tape for measuring corals and other features of the reef

Dive slates with waterproof paper and attached aqua pencils for recording data

Underwater digital cameras for documenting the ocean bottom and identifying species

Yellow mesh catch bag (visible on Randi Rotjan) containing a hammer, chisel, snips, and sampling bags

Yellow metal canister (visible on Rotjan, but all divers carried them) containing safety gear: an EPIRB unit (Emergency Position Indicating Radio Beacon) and VHF radio with a range of 5–10 miles

Divers (*left to right*) Stuart Sandin, David Obura, Les Kaufman, and Randi Rotjan

The damselfish that nearly got away: Gerry Allen's new species, *Chrysiptera albata*. (Gerald Allen)

found, so Gerry set off in search of another. With only a couple of minutes of bottom time remaining, we finally managed to coordinate our efforts and captured three specimens.

Greg and Austen had just returned to the boat after scouting for sharks when we broke the surface. Treading water next to them, Gerry peeled off his mask and yelled, "We got a new species!"

Our unprepossessing prize, now resting in a plastic bag on the skiff's pontoon, was a new deepwater damselfish species, pure white, without any distinctive markings, and the size of a business card. By damselfish standards, the new species would never win a beauty contest. But to an ichthyologist, this find was pure gold.

Bouncing off waves as the skiff motored back to *Nai'a*, Bailey recounted how we had seen several thousand surgeonfish and over five hundred bumphead parrotfish on our dive, describing the way the parrotfish had charged in a herd and grazed on corals, just like buffalo. We all voiced the hope that the fish in the Phoenix Islands wouldn't share the buffalo's fate. Along with the extraordinarily abundant fish, we found that the corals were also in a wonderful state: ninety-two species of live coral covered as much as three-fourths of the seafloor, and where there were no corals, we found healthy coverings of *Halimeda* and other calcareous algae, all of which indicated a healthy mid-Pacific coral reef community, providing abundant niches and vital habitats for a wealth of marine organisms. We eventually described the new damselfish species in a scientific journal, bestowing on it the scientific name *Chrysiptera albata*. The specific name *albata* means "clothed in white."

After 26 days and some 163 hours of diving, we had recorded a total of 438 species, including 209 new records. The reef fish fauna of the Phoenix Islands now stands at 509 species in 216 genera and 67 families. In addition to the new damselfish, we collected a new soldierfish (since described as *Myripristis earlei*) and several goby species that still await description. The Phoenix Islands inventory provides a vital piece in the fish diversity puzzle of the vast Indo-Pacific region, but there is still much to accomplish.

Gerry's work, outside of the Phoenix Islands expedition, is centered around what is called the "Coral Triangle": the water between Indonesia, the Philippines, and Papua New Guinea. Universally acclaimed as the world's richest location for marine life, Indonesia boasts 2,057 species of reef fish. Numerous studies show that the farther one travels from this region, the fewer species are present. Although the Phoenix archipelago has only about one-fourth as many reef fishes, it is an extraordinary place that is certainly worthy of special conservation attention.

These islands are perhaps the best example of a near-pristine atoll environment that we've had the pleasure to experience. We marveled at the bountiful cavalcade of undersea life on each and every dive.

4 BIRDS AND INVADERS
RAWAKI RAYMOND J. PIERCE

As we approached Rawaki from the south one evening in 2006, vast plumes of birds—thousands of frigatebirds, boobies, terns, and noddies—were streaming toward this small island to roost for the night. Later, in the darkness, we could hear a cacophony of calls coming from the invisible island nearby, and there was a pungent aroma of guano in the air.

Landing on treeless Rawaki the following morning was breathtaking, like stepping back a few thousand years to an era when seabirds were abundant throughout the Pacific. This arid, 124-acre (50 ha) island was alive with large colonies of sooty and grey-backed terns and brown noddies. Hundreds of noisy parent birds overhead protested our intrusion. Tiny blue noddies fluttered around our heads, some returning to their nests under the scant cover of *Portulaca* shrubs. Nesting masked and brown boobies were dotted conspicuously

The evening fly-on at
Rawaki. (Paul Nicklen)

Birds nesting in a wall of coral blocks: brown noddy adult and juvenile (*above*) and black noddy (*below*). (Ray Pierce)

around the island, and several pairs of red-footed boobies occupied coral slabs and other debris at the edge of the small lagoon. We estimated that there were some twenty thousand incubating and roosting individuals present in a tightly packed colony of lesser frigatebirds, and to one side was a less crowded colony of great frigatebirds. Several males of both species had fully inflated bright red gular (throat) pouches, which indicated that they were breeding. Above the high-tide zone, tucked beneath shady slabs of coral and scattered driftwood, were red-tailed tropicbirds—adults sitting on eggs or brooding well-fed chicks.

Frigatebirds nesting in regenerating *Sida* bushes on Rawaki. (Ray Pierce)

White terns. (Larry Madin)

Frigatebirds in flight, Rawaki. (Ray Pierce)

These birds were merely the day shift. We wanted to know whether some of the more sensitive and threatened bird species were also using the island, as that information would reveal the true ecological importance of Rawaki for seabird conservation. To find out, Aobure Teatata, from the Kiribati Wildlife Conservation Unit, and I conducted "fly-on" surveys: evening observations of birds returning to the island. Each survey entailed ninety minutes of scanning from a vantage point on the sandy upper beach. We needed to take care to avoid damaging any turtle nests, and we tried to keep the dipping sun over our right shoulders in order to get the best lighting for identifications. We stood watch along a 219-yard (200 m) corridor of the beach as thousands upon thousands of sooty terns, noddies, boobies, and other seabirds streamed in to the island amid a deafening crescendo of various calls, at a volume neither of us had heard before anywhere in our travels. Occasionally a frigatebird would swoop down to chase an unsuspecting returning red-footed booby and force it to disgorge its catch.

When searching for specific rarities amid the action and noise of this number of birds, one has to develop an effective means of filtering out the common birds—looking for the "jizz," or characteristic form and flight motion, of the target species. Suddenly, among the tens of thousands of incoming terns, Teatata and I spotted a single endangered Phoenix petrel gliding in wide semicircles on stiff wings. It was soon followed by others. Later, in the fading light of the evening, we started to see different shearwater species coming in with rapid wing-

beats, along with Bulwer's petrels. As darkness descended, we were excited to see increasing numbers of another endangered species, the delicate white-throated storm-petrel, aptly named *te bwebwe ni marawa*, or "sea butterfly," in Kiribati, fluttering slowly toward land. All these fly-on observations confirmed the importance of Rawaki to rare seabirds. They also confirmed that, although rabbits were present, there were clearly no rats or cats on the island; otherwise, these sensitive bird species would have been extirpated.

Teatata and I were part of a team of biologists studying how the Phoenix Islands' native birds and other flora and fauna were being affected by invasive species. Our study, which was funded by Conservation International's Critical Ecosystem Partnership Fund (CEPF), represented a first step by the government of Kiribati to restore the ecology of the islands.

By surveying the island at night—taking care not to stand on the multitudes of nests and burrows—we could further determine the status of these residents and sort out who was nesting where and in what numbers. After three days and nights, a clearer picture of the island's bird community emerged. Each species was occupying a different habitat, building a different type of nest scrape or burrow or, like the frigatebirds, nesting on exposed vegetation.

Audubon's and Christmas Island shearwaters returned to their nesting burrows in the sandy soil or beneath coral slabs, and we spotted several wedge-tailed shearwaters cleaning out larger burrows in the island's sandy interior. The Phoenix petrels and storm-petrels were concentrated in a narrow zone of surviving bunch grass (*Lepturus*), beneath which they had excavated their nest scrapes. As the birds returned to roost for the evening, we estimated that the total number of blue noddies climbed to some five thousand individuals.

On the lagoon edge, nearly a thousand masked boobies had formed a non-breeding roost on the baked mud, where they would spend the night. Other roosts of thousands of frigatebirds occupied areas of coral rubble, well removed from the main nesting colony of frigatebirds.

Near the lagoon, bristle-thighed curlews began to stalk hermit crabs into the night, using their bills to smash each crab's adopted shell against an anvil of coral in order to extract the crab. Clearly there were preferred anvil sites, as a few of the coral slabs were surrounded by hundreds of smashed shells. Pacific golden plovers foraged for other nocturnal invertebrates such as spiders, beetles, and small crabs, also gleaning seeds from the sparse herb fields. These and other shorebirds spend most of each year on the atolls of the Pacific, taking advantage of rich food supplies, but return to Alaska to breed during the brief Arctic summer.

Back at our camp (which our colleagues had erected while Teatata and I were seeing to observations that could not wait!), we reviewed an amazing day of observations. We reflected on and marveled at how Rawaki's prolific bird life has survived for millennia, offering a window into what many Pacific islands may have looked like before the arrival of man. The healthy marine environment of the region provides seabirds with rich sources of food, including fish, squid, and mollusks, while the inhospitable physical environment of Rawaki has effectively barred human settlement. The lack of a freshwater lens to supply drinking water and irrigation, a rocky coast that makes landing boats difficult (as we

A white-throated storm-petrel outside its nest burrow. Ground-nesting birds are most vulnerable to predation by cats and rats and to habitat changes brought about by over-browsing by rabbits. (Ray Pierce)

A Phoenix petrel at its heavily browsed nest site. (Ray Pierce)

A bristle-thighed curlew.
(Randi Rotjan)

found to our own dismay), and the island's classification as a bird sanctuary since the late 1970s—all these factors have contributed to Rawaki's survival as a bird refuge. Here the birds are able to roost and nest in relative peace.

The diversity and abundance of fish-eating birds at Rawaki also benefits other specialists, such as the two species of parasitic frigatebirds, also known as man-o-war birds. These large relatives of pelicans (the great frigatebird, *Fregata minor*, was originally thought to be a small pelican, *Pelecanus minor*) frequently pursue and harass boobies, noddies, and terns to force them to disgorge their food in flight, which the frigatebirds then snap up before the prey falls into the sea. Individuals of some species have adapted defensive strategies to avoid the frigatebird attacks, such as returning from foraging bouts after dark. Masked boobies, however, return to the islands in small groups of up to twenty in the early evening, but pause while still 150–300 feet (about 50–100 m) in the air, scanning the local area for the waiting frigatebirds. Then they fold their wings and hurtle down at great speed, occasionally pursued briefly by a frigatebird until the slower and less agile bird gives up the chase.

The colonies of frigatebirds and other seabirds also contribute to the food web with their guano, which provides nutrients for many organisms, both in the reef and in the open ocean. These organisms, in turn, provide food for reef and pelagic fish, ultimately sustaining the seabirds. While Rawaki is rich in guano resources, the other Phoenix Islands currently lack sufficient seabird concentrations to contribute significant nutrients to the food web. The restoration of other islands in the Phoenix Islands Protected Area (PIPA) should eventually spur the recovery of seabird populations and reinstate this nutrient recycling process.

Significance of the Phoenix Islands for Seabirds

The Phoenix Islands have been known for nearly fifty years to have internationally important seabird populations, including some of the world's largest breeding colonies. During the 1960s the Pacific Ocean Biological Survey Program recorded nineteen species of breeding seabirds in the Phoenix Islands, including colonies containing up to thirty thousand lesser frigatebirds. Today the same nineteen seabird species still breed within the PIPA. They span three orders and six families of birds: Procellariidae (petrels and shearwaters), Hydrobatidae (storm-petrels), Phaethontidae (tropicbirds), Sulidae (boobies), Fregatidae (frigatebirds), and Sternidae (terns).

Other nonbreeding visitors to all of the PIPA islands include about twenty shorebird species, four of them common: Pacific golden plover, wandering tattler, turnstone, and bristle-thighed curlew. One land bird (the long-tailed koel, a cuckoo) migrates to some of the forested islands from its breeding grounds in New Zealand. In addition, many other seabirds migrate in tens or hundreds of thousands through the Phoenix Islands en route to higher latitudes north or south of the equator.

Surveys conducted from 2006 to 2011 show that each island supports varying numbers and diversities of seabirds. Although Rawaki is the standout island in terms of sheer number and variety of birds, supporting nearly the entire suite of breeding seabirds in the Phoenix Islands, Orona hosted several spectacularly large colonies of sooty terns in 2006. These terns numbered more than the rest of the birds in the PIPA combined, but by 2009 their numbers appeared to have declined, probably as a result of predation on eggs and chicks by rats and on eggs, chicks, and adult birds by feral cats. The forested islands of Manra, Orona, and Nikumaroro tend to support more pairs of tree-nesting species—including the red-footed booby, black noddy, and white-tailed tropicbird—than islands without tree cover.

Today the numbers of birds are at the same order of magnitude as those recorded in the 1960s. This is remarkable, considering the sharp drop-off in numbers observed in many seabird colonies elsewhere in the Pacific.

The Phoenix Islands are globally important for seabirds because of the ongoing human-induced declines of seabirds around the world. The populations of several species within the PIPA—including those of Audubon's shearwater, Christmas Island shearwater, white-throated storm-petrel, lesser frigatebird, masked booby, brown booby, red-tailed tropicbird, grey-backed tern, and blue noddy—are among the largest recorded for their species. Moreover, some of these birds are threatened species—notably the Phoenix petrel and white-

Table 4.1: Estimated seabird breeding populations on the Phoenix Islands in 2006–11

SPECIES	KIRIBATI NAME	ENGLISH NAME	ESTIMATED TOTAL PAIRS
Pterodroma alba	Te ruru	Phoenix petrel	< 100
Bulweria bulwerii		Bulwer's petrel	< 50
Puffinus pacificus	Te tangiuoua	Wedge-tailed shearwater	500+
Puffinus nativitatis	Te tinebu	Christmas Island shearwater	500+
Puffinus lherminieri	Te nna	Audubon's shearwater	1,000+
Nesofregetta fuliginosa	Te bwebwe ni marawa	White-throated storm-petrel	100+
Phaethon rubricauda	Te taake	Red-tailed tropicbird	2,500+
Phaethon lepturus	Te gnutu	White-tailed tropicbird	10
Sula dactylatra	Te mouakena	Masked booby	5,000+
Sula leucogaster	Te kibwi	Brown booby	250+
Sula sula	Te koota	Red-footed booby	2,500+
Fregata minor	Te eitei are e bubura	Great frigatebird	1,250+
Fregata ariel	Te eitei are e aki rangi ni bubura	Lesser frigatebird	20,000+
Sterna lunata	Te tarangongo	Grey-backed tern	5,000+
Sterna fuscata	Te keeu	Sooty tern	1,000,000
Anous stolidus	Te io	Brown noddy	10,000+
Anous minutus	Te mangikiri	Black noddy	5,000+
Procelsterna cerulea	Te raurau	Blue-grey (blue) noddy	2,500+
Gygis alba	Te matawa	White tern	1,000

throated storm-petrel (both listed by IUCN as endangered)—that have virtually no secure populations elsewhere because of the impacts of introduced predators, particularly cats and rats. The Phoenix Islands are also important as a wintering area and "refueling stop" for the bristle-thighed curlew (listed as vulnerable), which breeds in Alaska.

The Impact of Introduced Pests

Despite its success as a seabird colony, Rawaki has not been spared the impacts of invasive species. European rabbits were released there to provide a source of meat for workers during the guano-collecting days of the 1860s and 1870s. The rabbits not only smash seabird eggs (in part to obtain moisture) and trample chicks, but have also had a deleterious effect on the island's ecosystem. In 2006 only four plant species were common on the island, and severe browsing by rabbits on surviving plants, such as *Boerhavia* and *Portulaca*, had greatly reduced the nest sites available to petrels, shearwaters, storm-petrels, blue noddies, and other bird species.

During our 2006 survey, we saw that large sections of the island had been converted to featureless desert sand by rabbit overbrowsing, robbing small seabirds of nest sites. Mike Thorsen, our invasives specialist from New Zealand's Department of Conservation, found that seedlings of the small shrub *Sida fallax* (kaura) were being browsed to oblivion before they could mature.

The other Phoenix Islands are all infested with rats, and most with feral cats as well. These two species have had more serious negative impacts on the birds than have the rabbits. One of the big disappointments during our 2006 survey came as we approached McKean Island. The vast numbers and variety of birds

SIDEBAR 4.1: The Strange Case of the Catchbird Tree · *Raymond J. Pierce*

(Oceanwide Images)

The seeds of trees in the genus *Pisonia* are very sticky, so sticky that the common name of these trees in the Pacific is the catchbird tree. *Pisonia grandis*— also known in some places as the mapou tree—is particularly common on small Pacific islands with large colonies of nesting seabirds, and it is common in the Phoenix Islands. It is a prolific producer of sticky clusters of up to two hundred seeds, whose sticky resin is particularly good at sticking to feathers. It's so sticky, in fact, that the seeds can actually immobilize birds and kill them. Is killing the bird a "deliberate" strategy of the tree, ensuring a bird carcass to fertilize its growing saplings? Or is it just a random consequence of the extreme stickiness of the resin in which the seeds are embedded?

Canadian researcher Alan Berger studied the problem, trying to determine whether mapou trees got any benefit when entangled birds died. On Cousin Island in the Seychelles, in the Indian Ocean, he conducted experiments with thousands of *Pisonia* seeds. He found that the opposite was the case: the trees did better when the bird lived to disperse the *Pisonia* seeds when it preened the resin from its feathers.

The mapou tree seems to have evolved sticky resin so that its seeds can withstand multiple dunkings in seawater. The hundreds of seabirds that die a sticky death each year appear to be collateral damage in the mapou tree's campaign to colonize and spread to other parts of the island and from island to island.

Sources: Journal of Tropical Ecology 21 (2005): 263–71; Nature Seychelles, accessed January 26, 2011, www.natureseychelles.org.

that the 1960s researchers had found there, and that we were hoping would still be present, were all but gone. Even before we landed, the absence of wheeling flocks of terns and frigatebirds was a telltale sign, and the recently wrecked fishing vessel on the beach provided a clue to the likely cause of the decline. On landing, we found that the island was infested with a large and aggressive rat that none of us had seen before. Genetic tests proved it to be an Asian rat (*Rattus tanezumi*) of the Korea Peninsula; it was probably no coincidence that the wrecked boat was from Korea.

Three other islands—Birnie, McKean, and Enderbury—are superficially similar in habitat to Rawaki, but Enderbury is about ten times larger. While all three are arid and have a central supersaline lagoon, McKean and especially Enderbury support much more diverse vegetation than is found on Rawaki. Mike Thorsen and Tiare Etei, from Kiribati's Ministry of Environment, Lands and Agricultural Development, documented a variety of grasses, herbs, and vines as well as shrubs and small trees on McKean and Enderbury. While this greater habitat diversity makes these two islands potentially highly attractive to breeding and visiting birds, we found that both islands, as well as Birnie, were infested with rats. Pacific rats were present on Enderbury and Birnie, while the Asian rat was present on McKean. The rats have greatly depleted the numbers and diversity of seabirds and other faunal groups, such as lizards and invertebrates.

The remaining four islands in the Phoenix group—Kanton, Manra, Orona, and Nikumaroro—are large forested islands and therefore support populations of some tree- and shrub-nesting seabirds, especially black noddies and red-footed boobies. Kanton supports the only other Phoenix petrel colony present in the PIPA, while Manra and Nikumaroro support modest numbers of a tree-nesting species, the white-tailed tropicbird. However, these four islands have also had a long history of human and predator impacts; rats are still present on all and feral cats on three. Consequently, these islands have seen reductions in most of their very sensitive seabird species.

Removing the Invaders

By the time we completed our first survey in 2006, it was clear that the PIPA's fragile terrestrial ecosystems and seabirds needed help. The islands can be restored only by eradicating the invasive rabbits, rats, and feral cats and by implementing effective biosecurity measures to ensure that McKean-style disasters never happen again.

In 2006, 2008, 2009, and 2011, we surveyed invasive mammal species on the Phoenix Islands. In 2008 and 2011, we undertook measures to eradicate these species to safeguard the islands as world-class nesting sites for endangered birds:

Rabbit (*Oryctolagus cuniculus*): Successfully eradicated from Rawaki in 2008.

Feral domestic cat (*Felis catus*): Present on Kanton, Orona, and Manra; eradication not yet attempted.

SIDEBAR 4.2: Disaster on McKean · *Raymond J. Pierce*

(Larry Madin)

During the 1960s, the Pacific Ocean Biological Survey Program recorded a high diversity and abundance of seabirds on Rawaki and McKean. At that time, both islands supported large numbers of storm-petrels, shearwaters, blue noddies, terns, and other sensitive species. As we approached McKean forty years later in 2006, we could see that something had gone badly wrong. No clouds of birds greeted us: the island was silent. Then we saw the remains of the merchant vessel *Chance*, a Korean fishing boat that had recently been wrecked on the McKean reef, providing easy access to the island for invasive stowaways.

On landing at McKean, we found that nearly all of the storm-petrels and blue noddies were gone, along with most of the shearwaters and terns. Only the large seabirds (boobies, red-tailed tropicbirds, and lesser frigatebirds) were hanging on. Large rats (the Asian rat, *Rattus tanezumi*, originating from the Korea Peninsula) had overrun the island and decimated the smaller birds. The bird populations may have already been in decline due to the presence of a smaller rat (probably the Polynesian rat, *R. exulans*) that had been reported here on some late-twentieth-century trips, but the Asian rat rapidly completed the extirpation of the small seabird species on the island. It took a special NZAID expedition in 2008 to successfully remove the rats from McKean, thereby making the island safe for all nesting seabirds once again.

Table 4.2: Impacts of invasive species on seabird breeding populations

	RABBIT	PACIFIC RAT	ASIAN RAT	CAT
Low to moderate impact	Most nesting birds	Boobies, tropicbirds, and frigatebirds	Larger nesting birds, such as boobies, tropicbirds, and frigatebirds	Tree nesters such as the red-footed booby, black noddy, and white tern
Serious impact	Petrels and shearwaters (Procellariidae), especially the storm-petrel and blue noddy	Eggs and nestlings of terns, shearwaters, and noddies		Ground-nesting birds
Catastrophic impact		Storm-petrel and blue noddy	Smaller nesting birds, including storm-petrel, blue noddy, all terns and noddies	Petrels and shearwaters (Procellariidae), storm-petrel, brown and masked boobies, frigatebirds, tropicbirds, blue and brown noddies, and other terns

Pacific rat (*Rattus exulans*): Present on Enderbury, Orona, Birnie, Nikumaroro, Kanton, and probably Manra. Eradication was attempted at Enderbury and Birnie in July 2011, and its success should be known sometime in 2012 or soon after.

Asian rat (*Rattus tanezumi*): Successfully eradicated from McKean in 2008.

Black rat (*Rattus rattus*): Present on Kanton and Manra; eradication not yet attempted.

The 2006 conservation survey rated Rawaki and McKean as most urgently in need of pest eradication to secure seabird populations and relieve the island ecosystems of chronic pressure from rabbits and rats. These top-priority eradications were undertaken by Pacific Expeditions Ltd. in May and June 2008. They were funded by NZAID (New Zealand's international aid and development program) and supported by several agencies, including Conservation International, Eco Oceania Ltd., the New England Aquarium, the New Zealand Department of Conservation, Pacific Expeditions Ltd., and the Pacific Invasives Initiative, as well as the government of Kiribati.

In November and December 2009, a CEPF- and NZAID-funded expedition visited the PIPA and declared Rawaki and McKean to be pest-free. The vegetation and bird populations of both islands showed clear signs of recovery. On McKean, for example, grey-backed terns and brown and black noddies were nesting very successfully, whereas during our previous visits in 2006 and 2008 most nests were failing. On Rawaki, blue noddies and other species were responding to increased vegetation cover by nesting in the former "desert" area of the island. We noted that frigatebirds were now nesting in *Sida* bushes, which previously had been suppressed by rabbits.

The PIPA Management Plan calls for eradicating invasive species from all of the Phoenix Islands and instituting an effective biosecurity plan to keep invasive species out. The highest priorities for restoration include Enderbury and Kanton. The proximity of Enderbury and Birnie to Rawaki (62 miles [100 km]

New Zealand hunter Lance Cooper fine-tuning a pointer's nose for rabbit detection during a 2008 eradication mission to Rawaki. (Ray Pierce)

each) means that once rats are removed, these islands will begin to be recolonized by seabirds from Rawaki. Indeed, during our 2006 and 2009 surveys of Enderbury, we were encouraged to see small numbers of Phoenix petrels, blue noddies, and shearwaters visiting the island, but they will not be able to establish a robust breeding colony until the rats and feral cats are removed. Enderbury is an arid island; it has a series of small ephemeral supersaline lagoons that are virtually devoid of animals, but the small islets in the lagoon provide important refuges for nesting brown noddies, safe from the teeth of rats. Elsewhere, isolated stands of trees provide nest sites for thousands of black noddies and hundreds of red-footed boobies. Enderbury already supports thousands of frigatebirds and masked boobies and hundreds of brown boobies. It is hoped that the smaller and rarer seabirds will successfully reestablish a presence on the island once the rats are gone.

Kanton is the gateway to the PIPA, and visits there will steadily increase with ecotourism. The island has considerable strategic value, given that a small colony of Phoenix petrels is hanging on by a thread there, and that there are some large colonies of terns and noddies on the islets at the entrance to the lagoon. Rats, cats, and invasive plants represent a threat not only to Kanton's habitats and species, but also to other islands if they stow away on one of the many vessels that visit Kanton. Removing invasives from Kanton will enhance biosecurity throughout the Phoenix Islands.

In 2011 an operation was undertaken to eradicate rats from Enderbury and Birnie, funded by the Packard Foundation, Conservation International, and the CEPF and supported by the other agencies previously involved with Rawaki and McKean. We will not know for a year or more whether these 2011 operations have been successful, but if the islands are pest-free, it will mean four down and four to go.

In order to rid the remaining four islands within the PIPA of invasive mammals, the cats and rats that occur there will have to be addressed. In contrast to the northern islands, the three southern islands of Manra, Nikumaroro, and Orona have more forested and scrubby habitat (and coconut plantations, established in the 1930s and 1940s or earlier), reflecting their higher rainfall and ephemeral freshwater table. Orona and Nikumaroro each have a large, beautiful lagoon. All three islands support a variety of lizards and invertebrates, including varying numbers of coconut crabs (*Birgus latro*) and crabs in the genera *Cardisoma* and *Coenobita*. The islands have excellent potential for restoration, but the high densities of large land crabs on Orona and Nikumaroro make rat eradication a challenge there, since some land crabs will eat rat bait. Although it does not harm them, they can remove large quantities of bait intended for rats. The poison can also accumulate in the bodies of shorebirds and waders that eat crabs, so appropriate precautions need to be taken.

Island Biosecurity

The success of ecological restoration depends on effective biosecurity measures to ensure that invasive species are kept off the islands in the future. The greatest risk is posed by rodents and pest invertebrates (especially non-native ants), which infest many of the vessels visiting the central Pacific. Any careless landing by pest-infested vessels will spell disaster for the islands and their biota, as happened with the rat invasion at McKean. Invasive ants can also prevent successful nesting by many seabird species and so could effectively cancel out any progress gained by eradicating rabbits, cats, and rats.

Biosecurity requires a three-pronged approach to preventing pest invasion:

1. Quarantine. This approach involves preventative measures at each visiting vessel's port of origin as well as on the boats themselves. For the islands of the PIPA, it means preventing illegal landings and eradicating rats, non-native ants, and other invasives from cargo and fishing boats.
2. Surveillance. This approach involves checking the vessels that pass through the PIPA as well as periodically checking the islands for any sign of pest re-invasion, grounding, or illegal landing.
3. Contingency response. This approach involves responding to the presence of an invasive species, such as a cat or a rat, in the PIPA and taking immediate action to eradicate it.

If these steps are taken, then the rich marine food resource and safe nesting sites provided by the islands' remote location will probably spark a recovery of seabirds, as it has after pest removal on many islands around the world.

Biosecurity depends on effective quarantine measures. Beginning in 2010 all vessels visiting the PIPA waters must be inspected and verified as strictly pest-free, whether landing or anchored offshore. Landings obviously heighten biosecurity risks, and only those necessary for essential conservation and research work are permitted. All expeditions must follow stringent biosecurity protocols at their port of origin, en route to the PIPA, and before landing at the islands. Such measures include vessel fumigation, control of invasives at wharves at departure ports, checking of all equipment and provisions being brought on board

(Ian Sarnat/PIA Key)

The risk of species invasions on Pacific islands is not limited to the rats or cats that swim or leap ashore from wrecked ships. Even if all invasive vertebrate species were eradicated from the Phoenix Islands, other risks would remain in the form of small invasives, especially ants.

Seabirds can lose eggs and chicks to invasive ants. These ants can take a surprising toll on nesting birds' success and can interfere with birds' survival in other ways as well. A study conducted on Christmas Island (Indian Ocean) showed that fruit-eating birds could be driven from their food source when they had to compete for fruit with aggressive yellow crazy ants (*Anoplolepis gracilipes*). These ants are widespread in the Pacific. They are often present at seaports and can easily stow away on vessels. Therefore, they represent an ongoing threat to plant and animal life on islands throughout the Pacific.

In the United States, where a refuge was established in Mississippi in 1975 to help protect the sandhill crane, the cranes are under a new threat from invasive fire ants. Not only do the ants compete with cranes for nest space, but these voracious predators can also devour crane eggs and even chicks. In 2009 the Mississippi Sandhill Crane National Wildlife Refuge began using volunteers to monitor the progress of another aggressive ant invader, the raspberry crazy ant (*Paratrechina* spp.), introduced through the Port of Houston sometime around 2002. The lesson for the Phoenix Islands is that biosecurity depends on the strict hygiene of all visiting vessels.

the vessel, maintaining rodent bait stations on board, carrying out pest surveillance on board, and if landing is permitted, thorough checking of all material to be taken ashore. Through its staff and representatives in Tarawa, Kanton, Kiritimati, and other inhabited islands, the government of Kiribati is committed to making these biosecurity measures work, but more resources are needed to make them effective, particularly at Kanton. Outside agencies, including Eco Oceania Ltd., are providing advice and other support.

The tide of invasive species has been slowed at the PIPA, and there is hope that it may soon be entirely halted, allowing the remaining native plants and animals to recover. But even with new biosecurity, the terrestrial flora and fauna of the Phoenix Islands will have conspicuous gaps. Land birds, in particular, have disappeared because of habitat change and decades of predation by humans and introduced predators. Restoration planners will have to consider which species are appropriate candidates for reintroduction to restore the ecosystem to what it was like in the 1880s.

Finally, the restoration of the Phoenix Islands cannot be considered in isolation from restoration in the central Pacific. In a worst-case scenario, climate change, ocean acidification, and rising sea levels may exert increased pressure on local seabird species. But for now, increasing the productivity and abundance of threatened and sensitive species like the Phoenix petrel, white-throated storm-petrel, and blue noddy will increase opportunities for their dispersal (or translocation) from the PIPA to other restored islands, thereby increasing their chances for long-term survival in the Pacific.

We can't know what the future holds for the Phoenix Islands and their birds, but we can take steps to help ensure that they thrive in abundance for generations to come—the birds' generations and our own.

Return to Rawaki

In July 2011 the eradication team made a fourth visit to Rawaki, one of the first two islands in the Phoenix Islands to be declared pest-free.

Accompanied by Derek Brown and Graham Wragg, key players in the removal of rabbits from the island in 2008, we aimed to collect information on the responses of the plants and birds three years on from the removal of the invasive browser.

Six of us had briefly visited the island in November 2009 to confirm that the rabbits were indeed gone. Nevertheless, it was with some tension, then relief, that we returned to the sites where Derek had shot the last animals three years ago, checking en route for signs of browsing and, as night approached, any sign of rabbit movement. There was nothing to suggest any survivors, though, and all the rabbitlike silhouettes we saw were indeed juvenile boobies! The island was pest-free.

We returned to the photo-points (fixed sites for photographing changes in vegetation) set up in 2008 and could immediately see changes to the landscape. There was more and taller vegetation, particularly *Portulaca* and *Boerhavia*. A striking change was the many stands of *Sida fallax* that we now encountered; this woody shrub had been browsed to near-oblivion by the rabbits. These changes seemed all the more remarkable to us given the very dry conditions prevailing on the island in 2011.

Other obvious changes since 2008 included birds taking advantage of the new vegetation for nest sites. *Sida* was now the preferred nest site for frigatebirds; we observed colonies of both great and lesser frigatebirds nesting in *Sida* for the extra elevation it provided. In places we saw *Sida* apartment blocks, with frigatebirds occupying the top floor and red-tailed tropicbirds directly below on the shaded ground floor!

During this same visit, Andrew MacDonald, Katareti Taabu, and I completed an evening seabird fly-on survey and detected the full suite of seabirds present at Rawaki, including many Phoenix petrels and white-throated storm-petrels. At night we completed some targeted surveys of colonies (selecting sites to avoid disturbing nesting frigatebirds while including some of the previously denuded sites now supporting vegetation). Immediately we could see that blue noddies, shearwaters, and storm-petrels were nesting in or frequenting areas farther across the island than had been the case during the rabbit infestation. Instead of being confined to the shade of coral slabs and the limited *Lepturus*, *Sesuvium*, and *Portulaca* stands that existed before 2008, all of these birds were now more widespread, being able to nest in more prolific stands of *Portulaca* and *Boerhavia*.

We had limited time to quantify more subtle changes. Now predation on seabirds' eggs appeared to be confined to bristle-thighed curlews and crabs; no longer was there any sign of eggs being smashed for moisture, as rabbits had been doing. The increased plant cover was also providing more extensive habitat for invertebrates and therefore greater shorebird feeding areas.

Subsequent comments made by crew members on our ship, the M.V. *Aquila*, indicated that the visit to Rawaki was the highlight of their trip. While a good range of seabirds were seen on the other islands visited, the sheer diversity, numbers, and noise on Rawaki drove home to them the benefits that could accrue from making the other islands pest-free.

Expedition Diary: A Survey of Manra, 2011

RAYMOND J. PIERCE

In July 2011, six years after our first attempt, we finally made a successful landing at Manra to assess whether invasive pests could feasibly be eradicated.

Previously we had not been game enough to attempt a landing through the pounding surf of the coral reefs. This time, however, we had the luxury of two helicopters. Taking off from the deck of the M.V. *Aquila*, we had been using the copters to carry out rat eradication work on Enderbury and Birnie.

Once on Manra, we made our camp beneath a small grove of coconut trees on the edge of the abandoned village. We saw many relics, including a century-old concrete water tank as well as many wells, building foundations, and fences, all intricately built from coral slabs. In the middle of this area, Andrew MacDonald found the wreckage of an aircraft, apparently all that is left of a DC3 that crashed on Manra in 1943, killing nine U.S. airmen. According to the villagers' account at the time, the plane had clipped a coconut tree before crashing. It is likely that bits of this aircraft have turned up as traded items across the southern Phoenix Islands, which were also settled in the 1940s.

We made five forays across the island. One team rode in the helicopter to take aerial photographs of the entire island and visit islets in the supersaline lagoon; Graham completed a survey of the outer island's foreshore; Katareti and I surveyed birds along the lagoon edge and in the indigenous forest; Derek Brown, Andrew, and Nick Torr set rat traps and surveyed for cats in the coconut plantation; and Stacie Hathaway surveyed for lizards.

Our findings were intriguing. The rat traps produced only black rats (*Rattus rattus*), although it is possible that Pacific rats (*R. exulans*) also occur in the indigenous vegetation, as both species had been reported on Manra in the past. We found cat feces and skeletons in several spots, centered on the old village. The visits to the lagoon islets quickly revealed large colonies of successfully nesting terns and noddies, indicating that the predatory mammals seldom if ever crossed to the islets.

Graham's perimeter survey revealed about a thousand red-tailed tropicbird nests centered at Manra's eastern end, possibly a reflection of the lack, or near lack, of cats at that more arid end of the island. The behavior of the masked boobies along the lagoon edge was revealing: at our approach, both adults and juveniles adopted an aggressive threat display—clearly an adaptation to dealing with cats. Unfortunately, other smaller birds in the Phoenix Islands are unable to intimidate cats in this way.

Returning to camp, Katareti and I spotted white-tailed tropicbirds, confirming Manra as one of only two islands in the PIPA (the other being Nikumaroro) where this tree-nesting species occurs. In the evening we completed a fly-on survey that revealed all the seabird species using the island. As anticipated, we saw no shearwaters, petrels, or storm-petrels, all species that are susceptible to cat and rat predation.

The lagoon edge was also fascinating from a historical perspective. The coral foundations of nineteenth-century railway lines used for collecting guano were perfectly intact. We also found many guano test pits still in good condition.

Stacie's surveys revealed oceanic geckos (*Gehyra oceanica*) and mourning geckos (*Lepidodactylus lugubris*) in good numbers and, pleasingly, no introduced house geckos (*Hemidactylus frenatus*).

Manra is a relatively large island with a range of habitats and areas of indigenous forest vegetation, and its potential for the recovery and restoration of seabirds and other indigenous species (including the possible reintroduction of land birds) is considered to be high. The information gathered from the Manra survey will be used to develop pest eradication plans and budgets for further restoration work within the PIPA.

5 BEAUTIFUL ALIENS
THE INVERTEBRATES

MARY JANE ADAMS, M.D.

I joined the 2002 expedition to the Phoenix Islands as the medical officer and a member of the scientific team. Unlike most members of the expedition, I don't have a Ph.D. in marine biology; I participated as a citizen scientist. I am a medical doctor, retired from thirty-five years of practicing anesthesiology in a community hospital in Southern California. I am also an experienced scuba diver and underwater photographer with an avid interest in tropical marine invertebrates and fishes.

I had previously logged eighty-eight dive trips in the tropical Indo-Pacific in such prime diving sites as Indonesia, Papua New Guinea, and the Philippines—the famous Coral Triangle—but my dives in the Phoenix Islands were without a doubt some of the most exciting and memorable of my twenty-seven years of exploring underwater.

Since we were the first scuba divers ever to explore the waters of the entire archipelago, every time we rolled backward into the sea, we were going where no human had gone before. Very few people in the twenty-first century will ever experience this same thrill. I thought to myself, *Who says outer space is the last frontier?*

On the 2002 expedition, our survey team dove selected sites around each island and within the lagoons of Kanton and Orona. Conditions varied from pleasantly calm to rock-and-roll surge and screaming currents. The windward sides of the islands presented the most difficult conditions, but all of the Phoenix Islands are so far away from any large landmass that even the leeward sides are not very well protected. Some shallow corners of Kanton's lagoon were not flushed during tidal changes and contained dirty water, slimy algae, and dead corals. Except for these occasional murky corners of the lagoons, the underwater visibility was exceptionally good.

Everyone on the scientific team had a different job. The ichthyologists had their own protocol for surveying the fishes on each reef; David Obura had a different dive plan for measuring and counting hard coral species. I usually took off on my own to look for invertebrates, but sometimes helped the fish team

Mary Jane Adams on deck. (Cat Holloway)

As Greg Stone was planning for the second Phoenix Islands expedition in 2002, he realized that he would need to recruit a medical officer as a member of the team. Since the Phoenix Islands lie hundreds of miles from the nearest medical facility, the expedition team needed a ship's doctor. I had served as a science diver and photographer on the first Phoenix Islands expedition in 2000, and Greg approached me to see if I would be willing to take on the extra duties. Having recently retired from my anesthesiology practice, I was available for a six-week adventure, so I eagerly accepted his offer.

Medical problems are inevitable during a dive expedition to such a remote location. I brought along a well-stocked medical bag, but in case of a serious accident or illness, we needed an evacuation plan. Cruising back to Fiji from the Phoenix Islands would take five days—an eternity for a patient in critical condition. Samoa is a little closer, but transporting a patient to either place aboard *Nai'a* would mean that we would have to abort the whole mission. We knew from the first Phoenix Islands expedition that there was a World War II airstrip on Kanton. Although it hadn't been used in decades, it was still in good condition. So we worked out a plan for a charter flight from Samoa to Kanton in case of emergency. This would shorten the time to a hospital to about three days—the same as in the contingency plan for evacuating Apollo 11 astronauts from the moon. Fortunately, I never needed to call for outside help.

The most common complaint among the passengers on our voyage was seasickness. Almost everyone suffered some *mal de mer* during the five-day crossings between Fiji and the Phoenix Islands. I brought a large supply of nausea medication and used every bit of it.

The most worrisome problems I treated were infections. When skin is exposed to seawater for several hours a day, every little nick and scratch becomes infected, usually with staphylococci. If not treated early and aggressively with antibiotics, a tiny break in the skin can quickly develop into a serious infection, even septicemia. We used a lot of topical antibiotics during the expedition, and in a couple of cases oral antibiotics were required.

One of the risks of deepwater diving is decompression sickness, or the "bends." For the 2009 expedition, ship's doctor Craig Cook arranged for a portable recompression "stretcher" to be carried aboard *Nai'a*. These innovative lightweight hyperbaric chambers can be easily deployed in remote locations, allowing patients to be stabilized and begin treatment on the ship's deck before emergency evacuation. The chamber can be pressurized to a depth of 80 feet (approx. 25 m) using the compressed air in one scuba tank. Despite some extreme diving in unknown waters, no diver has experienced decompression sickness on any of the Phoenix Islands expeditions.

collect specimens after they had applied the chemical rotenone to the crevices in the coral rubble.

Rotenone is a plant-derived piscicide used to stun or kill fish. In the field it allows scientists to collect fishes that would otherwise be difficult to sample. On deck, Gerry Allen and Steve Bailey mixed the rotenone to a pancake batter–like consistency in a bucket, then scraped this batter into wide-mouthed plastic bags. The bags were then twisted shut and placed in a mesh dive bag for safe transport to the collection site on the bottom. Once there, Gerry and Bailey untwisted the bags, releasing the rotenone into the crevices of the coral rubble where the well-camouflaged fishes were hiding. Then I helped collect the specimens before the sharks and other predators could eat them.

By the end of the dive, our various assigned activities and the prevailing currents had often taken us a long way from where we had entered the water an hour earlier. Our lives depended on our keen-eyed skiff driver spotting us wherever we surfaced. Sometimes when I came up, I could see only fleeting glimpses of the skiff between swells. I had a few anxious moments bobbing around in the open ocean, wondering if the skiff driver would ever see me, but somehow he always did.

During the first Phoenix Islands expedition in 2000, my job was to photograph marine life and help the scientific team collect and process samples of fish and algae. On the second expedition in 2002, my duties expanded to include a survey of mobile marine invertebrates: mollusks, echinoderms, and crustaceans, as opposed to the sessile sponges, sea fans, and corals. Later, at the Natural History Museum of Los Angeles County, I worked with malacologists, the biologists who study mollusks, and other specialists to identify the specimens.

I found a large variety of crustaceans in the Phoenix Islands. Most were quite small, like this mantis shrimp (*pages 60–61*). (Mary Jane Adams)

Shelled and Shell-less Mollusks

My favorite subclass of mollusks is the Opisthobranchs, gastropods with reduced or absent shells, commonly called sea slugs. Sea slugs include nudibranchs, sea hares, side-gilled slugs, headshield slugs, and sap-sucking slugs (though the "sap" the latter suck is actually algae). Sea slugs are some of the most gaudy and colorful creatures in the sea, with a dizzying array of patterns and forms. The bright colors often serve as warning coloration, as they do in many insects and amphibians on land.

Nudibranch means "bare gill," and most nudibranchs breathe with a single dorsal gill. Sea slugs in the pleurobranch family, known as side-gilled slugs, have a hidden gill located between the mantle and the foot, usually only on their right side. Still other nudibranchs lack a gill altogether and breathe through their skin. The numerous soft, fingerlike appendages on the backs of these species may serve to increase the surface area available for respiration.

Nudibranchs tend to occupy the ecological niche underneath rocks and coral rubble, so I spent a fair amount of each dive digging around the bottom with my gloved hands. By doing so I found a wealth of bivalves, marine snails, large black brittle stars, sea stars, and sea cucumbers, in addition to a limited number

SIDEBAR 5.2: Nudibranchs

Chromodoris kuniei. (Mary Jane Adams)

Over evolutionary time, the shell-less mollusks known as nudibranchs have acquired an astonishing variety of hallucinogenic colors and shapes, reflected in such common names as orange gumdrop, shag rug, and fried egg. Their life histories are just as colorful.

To begin with, they're blind hermaphrodites, making their way through the environment by following chemical cues in the water. While they lack the classic molluscan mode of defense—a shell—they have developed two different strategies that make them exceptionally well-defended as they make their way over the reef surface.

One of the nudibranchs collected by Mary Jane Adams during the 2002 expedition was *Pteraeolidia ianthina*, known to divers as the blue dragon sea slug, which is widespread throughout the Pacific. Young animals start out colorless, but as they age, the soft spines on their body, known as cerata, take on colors ranging from tan, green, and brown to the bright blue that gives these animals their common name. The color comes from symbiotic algae in the tissues of the hydroids that are the nudibranch's preferred food. The algae accumulate in the cerata, where they provide their new host with the same nutritional boost they provided to the hydroid. But the benefits of a hydroid diet don't stop there. The nudibranch also manages to harvest the hydroid's nematocysts—its specialized stinging cells—and put them to use for its own defense.

A different group of nudibranchs, the dorids, has a different defense strategy. Instead of preying on hydroids, anemones, and other creatures with stinging cells, these nudibranchs feed on toxic sponges. They absorb the toxins in a feat scientists call sequestration, storing those toxins in their own bodies to make themselves unpalatable to predators. Pharmaceutical companies are studying the toxins in nudibranchs in search of new compounds that might yield drugs to combat cancer and other diseases. The nervous system of the nudibranch, with its outsize nerve cells and a relatively large brain for an invertebrate, is being studied to improve our understanding of the chemical basis of human memory.

of sea slugs. I placed all the specimens in plastic collection bags, labeled them with the time, date, and place found, and carried them to the surface in a mesh bag attached to my weight belt. In the end, I found fifteen species of sea slugs. Most of them were less than three-quarters of an inch long (2 cm), and a few were just over a third of an inch (1 cm). The only species I found in more than two locations was *Phyllidiella pustulosa*, one of the most common nudibranchs in the tropical Indo-Pacific.

I am accustomed to diving the Coral Triangle, a 1.6 billion-acre zone in the Pacific near Malaysia and Indonesia where marine invertebrate diversity is higher than anywhere else on Earth. By contrast, invertebrate diversity in the Phoenix Islands wasn't as high as in some other locations I had dived, though it was still far richer than other locations around the world, such as the Caribbean.

There may be several reasons for the relative lack of opisthobranch diversity in the Phoenix Islands. For one thing, the Phoenix Islands are extremely remote—1,000 miles (1,600 km) from Fiji. Most sea slugs and related species develop from an egg into a free-swimming larva, called a veliger. These veligers would have difficulty traversing great expanses of ocean, since they spend hours each day feeding on plankton. Another problem for traveling sea slugs is that they have very limited diets: each species of sea slug relies on only one or two species of sponges or soft corals for food.

By contrast, as I continued exploring the invertebrates of the islands, I found that the shelled mollusks—both bivalves and marine snails—were plentiful and sometimes appeared in staggering quantities. At Orona, the sandy bottom of the lagoon was dotted with giant clams. Near the channel leading to the sea, we found patches of reef completely overgrown with the giant clam *Tridacna squamosa*. This species normally grows to 16 inches or more (40 cm), but in this channel the clams were so crowded together that most of them had only reached about half that size. (A related species, *Tridacna gigas*, is a true giant, reaching as much as 4 feet [121 cm] across.) Such heavy concentrations are rarely found anywhere else. How and why these giant clams grew so plentifully in this channel remains a mystery: Are the conditions ideal for these clams, or did some long-ago settlers seed them in an early attempt at mariculture? The flesh of these huge bivalves is prized and eaten widely around the Indo-Pacific, and the shells may have been valued as raw material for personal implements such as combs and jewelry and as trade goods.

At Orona, near the channel leading to the sea, we found patches of reef completely overgrown with the giant clam *Tridacna squamosa*. (Paul Nicklen)

SIDEBAR 5.3: Micromollusks

A micromollusk is any adult mollusk with a shell between 0.4 and 0.079 centimeters—not much bigger than a large grain of sand. The scientists who study them have given these minuscule mollusks such scientific names as *Bittium* and *Ittibittium*. Under a magnifying glass, their tiny shells have all the details and structures, twists, and spirals of larger shells. Few researchers work on micromollusks because their size makes them especially difficult to study. Marta deMaintenon, a marine evolutionary biologist at the University of Hawaii, has taken up the challenge. She is studying dove shells—micromollusks in the family Columbelliae—in the hope of learning more about the evolution of extremely small body size. Biologists also want to know more about the biodiversity of these small marine organisms as well as those that live in ocean sediments. Mary Jane Adams collected more than a thousand mollusk specimens—itty-bitty and otherwise—that are now housed in the Natural History Museum of Los Angeles County, where they are being studied to increase our knowledge of mollusk diversity in the Pacific.

I also collected sediment from the base of reef walls for the Natural History Museum of Los Angeles County in the hope of finding micromollusks in the sand and grit. From 2 pounds (less than a kilo) of bottom sludge, the collections manager at the museum, Lindsey Groves, was able to extract 363 shells representing 55 genera and about 82 species.

Echinoderms

The most common sea star our team found in the Phoenix Islands was *Linckia multifora*, a colorful species sometimes called the Dalmatian *Linckia* for its variable red spots. These sea stars were abundant on the reefs and underneath the rubble on all the islands. This species is one of few that reproduce asexually by deliberately severing an arm in a process called autotomy. The amputated arm regenerates a new central disc and four arm buds that grow into a full-sized sea star. The original arm with the new arms growing from its base is sometimes called a comet sea star. The fact that individuals don't need to find a mate to reproduce may explain why *Linckia multifora* has been so successful while other sea stars in the Phoenix Islands appear relatively rare. I also found a couple of giant crown-of-thorns sea stars (*Acanthaster planci*) on Nikumaroro and single specimens of a few other species scattered around the islands. Although crown-of-thorns sea stars prey on hard coral polyps, the small number that we saw in the Phoenix Islands don't appear to pose a threat to the overall health of the reefs.

Sea cucumbers (holothurians) were scattered around the fringe reefs, but were most common in the lagoons. One edible species, the aptly named *Holothuria edulis* ("edible holothurian"), was abundant in Orona's lagoon and on sandy areas on some of the other islands.

The crown-of-thorns sea star (*Acanthaster planci*) is one of the most voracious predators of corals, but in low-nutrient and undisturbed conditions like those in the Phoenix Islands, it occurs at low-enough densities to coexist with healthy reef communities. (David Obura)

Sea urchins, both short-spined and long-spined, were rather scarce on the reefs, but I did find dense populations of an unidentified species near shore in some areas. Populations of sea urchins elsewhere ebb and flow as a result of changes in food abundance, disturbance of their habitat by storms, and disease. We still have more to learn about the role of sea urchins in the ecology of the Phoenix Islands reefs. Studies of the reef communities off the coast of Kenya suggest that sea urchin populations there are kept in balance by the presence of predators such as triggerfish.

The most difficult echinoderms to collect were the brittle stars, a fact I lamented during my dives. These ancient relatives of sea stars date back to the Ordovician, 505–440 million years ago. Most of the 1,500 species thus far described favor deep water, but brittle stars are known from reefs worldwide. Most of the brittle stars I found under the rubble were *Ophiocoma erinaceus*, large black brittle stars with long, thorny arms. To my frustration, they frequently slithered away through cracks faster than I could dig them out. Not only were they hard to catch, but it was difficult to get them out of the gravel intact because, true to their name, their arms broke off easily.

Interestingly, no one on the dive team saw any crinoids, also known as feather stars, on any of our island surveys. I don't know why they wouldn't occur in the Phoenix Islands. Crinoids aren't found at diving depths in Hawaii either, although they are plentiful on most other Pacific islands. I can only speculate that they have never successfully colonized these remote islands. Predation by crown-of-thorns sea stars or emperor fish seems a less likely explanation, though both are known to prey on crinoids elsewhere.

In total, we brought back 180 echinoderm specimens from the 2002 expedition.

Crustaceans, Polychaetes, and Flatworms

I found a large variety of crustaceans in the Phoenix Islands. Most were quite small, but a few dinner-plate-sized crabs and spiny lobsters peeked out of crevices in the reefs. Although polychaete (segmented) worms were not on the list of animals we hoped to survey, I collected the ones I encountered while searching for other invertebrates. Upon my return to Los Angeles, polychaete experts at the Natural History Museum of Los Angeles County identified seven different species from six families. No flatworms were observed during either expedition. I wasn't looking for flatworms, but it was still a little unusual not to see any.

Corals provide the red-spotted guard crab with a refuge from voracious wrasses. In return, the crab defends the coral colony against marauding predators such as the crown-of-thorns sea star. (Brian Skerry)

It's clear how these small, tube-building polychaete worms became known as Christmas tree worms. Their scientific name is *Spirobranchus giganteus*. (Jim Stringer)

A Rainbow Beauty

Finding any nudibranchs in these islands was a challenge. On some days of the 2002 expedition, I searched for hours without finding a single one. Any nudibranchs that were active during the day must have been well hidden in the crevices of the coral. By the time we reached Orona, the last island on our itinerary, I had only about a dozen species and had almost given up on finding a new one. After several dives on fringe reefs, we maneuvered the skiff through a narrow, winding channel into the lagoon. I expected to find seashells in this kind of habitat, but not sea slugs. The bottom was covered with mounds of snow-white sand that were occasionally interrupted by serpentine ridges of low reef. Thousands of giant *Tridacna* clams with iridescent blue, green, and purple mantles were crammed together on every available hard surface. Pink and black sea cucumbers lay around on the sand like carelessly discarded sausages. After collecting a few of the sea cucumbers, I started sifting through the sand in search of burrowing animals.

By the time my air pressure gauge reached the 500 pounds per square inch warning line, I had collected an assortment of shelled mollusks but hadn't found any sea slugs. Just as I started to ascend, I spotted a gorgeous multicolored nudibranch crawling over some dead corals. Stretched to the max, it was only half an inch (12 mm) long, but what a gem! I had never seen a sea slug like this before. The two tentacle-like organs on its head, called rhinophores, were amazing, with five pastel-colored bands that resembled tiny rainbows. Could this be a new species? I was so excited that I popped my rainbow beauty into a plastic bag and was halfway to the surface before I realized I had forgotten to photograph it.

Photographing invertebrates in their natural habitat is essential because they rapidly lose their color and shape once brought to the surface. In addition, invertebrate specimens are photographed in situ whenever possible to show the habitat they came from and, if feeding, their prey item. The photographic evidence also shows what the animal looked like before it was disturbed. Sea slugs go into spasm and retract their gill and rhinophores when disturbed. Even fishes change color rapidly after collection.

My 12-millimeter nudibranch had now scrunched up to about 8 millimeters and didn't want to come out of the bag. With only a few minutes of air left in my tank, I finally coaxed it back out onto the coral. I finished off my roll of film as fast as I could, fearing that my precious find would disappear into a crack and I would run out of air before I could get it out.

Back on board *Nai'a*, everyone wanted to know the identity of my spectacular little nudibranch. I searched through all of my books, but I couldn't find anything that looked like it. To satisfy our curiosity, Rob Barrel used his new satellite telephone to send a digital image of it to nudibranch experts in California.

The next day I received e-mail responses from Terry Gosliner at the California Academy of Sciences in San Francisco and Angel Valdes at the Natural History Museum of Los Angeles County, identifying my rainbow beauty as *Durvilledoris lemniscata*. Although rare, it was not a new species. It had been described 170 years ago by Quoy and Gaimard, ship's naturalists for the French research ves-

I was very excited to find this small dorid nudibranch in the white sand of Orona's lagoon. I didn't know what it was when I collected it, as it is quite uncommon, but got a positive ID via satellite e-mail within twenty-four hours of finding it. (Mary Jane Adams)

sel *Astrolabe*, on her voyage of discovery in French Polynesia in the 1830s. I was the first person to record it in the Phoenix Islands.

*

Back home in California, the scientific work on the invertebrate specimens began. As I reacclimated to my daily routine, Lindsey Groves, curator of mollusks at the Natural History Museum of Los Angeles County, was going through the shipment of specimens we had brought back from the Phoenix Islands.

After the first Phoenix Islands expedition in 2000, I was asked whether it was worth ten days of tossing around at sea to visit these tiny specks of land in the middle of the Pacific Ocean. My unequivocal response was "absolutely." There is something enchanting about these islands. Each one is unique, but they all seem to have a kind of mystical quality. Perhaps it is the excitement of discovery or of being one of the privileged few to walk under swaying palm trees and swim over virgin reefs on uninhabited atolls. Maybe it's just that being so far from "civilization" enabled me to fantasize about what Earth was like before humans arrived on the scene. Certainly, it was a more peaceful place. So, after two trips to the Phoenix Islands, would I do it again? Absolutely!

Expedition Diary: Diving the Pelagic Zone, Nikumaroro, 2009

GREGORY S. STONE

In September 2009 my colleagues and I dove the open waters off Nikumaroro. Accompanying me on this "blue-water" dive were Larry Madin and Kate Madin of the Woods Hole Oceanographic Institution and Alan Dynner, trustee and chair of the Board of Overseers of the New England Aquarium.

Blue-water diving takes place far from coral reefs and their colorful fauna. Such open water, called the pelagic zone, is home to the organisms Larry Madin has made his life's study: the tiny, gelatinous creatures called zooplankton. We had come in the hope of sampling them.

"Zooplankton" is the term for the tiny marine animals that drift in unimaginable numbers in the open ocean. Together with phytoplankton (tiny floating algae), bacteria, and other microscopic life forms, they form the base of the marine food web on which all life in the ocean ultimately depends. Zooplankton range from the familiar, like sea jellies, to lesser-known creatures like siphonophores, ctenophores, and salps, which are seldom seen by divers who stick close to the colorful, light-filled shallows of the coral reef. Soft-bodied, covered in mucus, often iridescent, these creatures appear as alien as beings from another world.

Blue-water divers are linked by safety lines to each other and to the dive skiff. (Jim Stringer)

Larry Madin sampling plankton during a blue-water dive in 2009. (Jim Stringer)

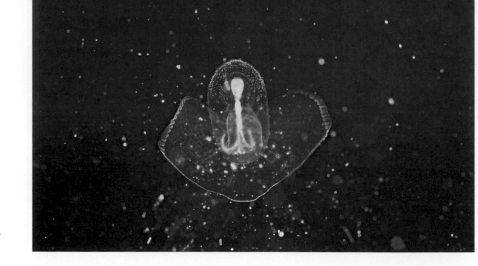

A swimming snail larva from a sample of plankton. (Brian Skerry)

And another world is what these deep blue waters resemble. In order to obtain a representative sample of pelagic plankton, we had to dive far from shore and well above the seafloor. On our 2009 dive, Nikumaroro was 4 miles away, and the ocean floor 10,000 feet below. The four of us descended, linked by safety lines to a 150-foot (46 m) line that connected us to the bobbing skiff. Below us the blue waters faded to a vertigo-inducing black. We relied on our safety lines to help keep us oriented and tethered to the skiff.

While my job on the dive was to observe and collect zooplankton, I had two additional assignments. As the designated safety diver, it was my responsibility to make sure no one's line came undone and allowed them to drift away. I was also the designated spotter for sharks that might come circling up out of the pelagic gloom. On this particular dive, I encountered one of the most aggressive sharks in my almost thirty-five years of diving and many thousands of dives, a 5- or 6-foot (1.5–1.8 m) gray reef shark that shot at me like a bullet as soon as I entered the water. I fended it off with my shark stick, causing it to retreat about 30 feet into the blue water, only to charge again, twice as fast this time and straight at me. I had only my shark stick between me and the animal. This time, it swam off and did not return. I have dived with sharks many hundreds of times, under all conditions, day and night. I have never had a shark so unexpectedly dart at me like this, and to this day, I have no idea why it did. But the gray reef shark left us alone for the remainder of the dive, to our great relief.

Most of the public's interest in the Phoenix Islands Protected Area is focused on the colorful coral reefs that surround each of the eight islands. But the coral reefs comprise only a tiny fraction of the PIPA. More than 99 percent of the ocean habitat in the marine protected area is in the pelagic zone. This zone is where a vast array of marine invertebrates live and drift; it's also home to schools of large oceangoing fish such as tuna and pods of whales and dolphins. Its depths, including seamounts and deep-sea creatures, still await exploration and discovery. Our blue-water dives mark the beginning of our attempt to understand this great frontier.

6 CORAL REEFS AND CLIMATE CHANGE

DAVID OBURA AND RANDI D. ROTJAN

Coral Castles DAVID OBURA

The story of the Phoenix Islands is largely the story of its coral—the reefs and atolls that make up this island chain. Coral has created a living barrier that for centuries has both attracted and repelled human settlers. The coral outcrops that make up the islands offer little or no fertile land for farming and little fresh water, and the small islands have almost no safe harbor for boats. At the same time, the reefs themselves provide vital food and shelter for a wide range of marine life that can support people, and the lagoons provide anchorage and protection from the vast blue expanse of the open ocean. The reefs serve as a way station not only for people, but also for large oceangoing fishes; the sheltered lagoons serve as a nursery for sharks and other species of fish.

The Phoenix Islands include both atolls and islands. The largest of the group—Kanton, Orona, and Nikumaroro—have fully formed lagoons. The Kanton lagoon is the largest and most spectacular, over 9 miles (15 km) long and 25 miles (over 40 km) square. With each tide, fresh seawater flows into and out of the lagoon through a single channel. The smallest islands vary in size from McKean, at 1.9 square miles (3 km²), to Manra, at 8.7 square miles (14 km²). While they do not possess true lagoons, all have central lagoonlike depressions with brackish ponds that support birds and other animals.

Each of the islands is uniquely oriented, swept by ocean currents and sheltered from winds in different places. These variations in the islands' interface with waves and winds create a variety of microenvironments, each exploited by a different reef community. On each dive, the scientists on the 2000 expedition found something new to marvel at.

Like most coral atolls, the Phoenix Islands have moderate species diversity, but relatively rare species occur there in great abundance. Vast areas of the outer reef are dominated by dense stands of the branching coral *Pocillopora*, the foot-long branches packed so closely together that they form a canopy like trees in a forest. Another classic feature of remote coral atolls is their surf-pounded coastlines. On the leeward sides of the islands, the sheltered coral communities

A healthy reef, replete with living corals and pink crustose coralline algae, which promotes new coral settlement. (Randi Rotjan)

resemble the picture-perfect images tourism boards use to lure divers to their sites. In contrast, the exposed southern, eastern, and northern sides tell a different story entirely: they are harsh, wild, and raw, pounded by massive waves and chiseled by ocean currents. Here the island slopes are covered in crustose coralline algae, which form a living pink carpet on which corals can gain a foothold. Some of the same corals grow on both windward and leeward sides, but on the windward sides the corals are pounded and broken into stunted forms, never reaching the sizes and development of the leeward sides.

On the platforms, reef edges, and deeper slopes of the sheltered western sides, we found luxuriant growths of corals and profusions of reef fish and invertebrates: sea fans, anemones, sponges, mollusks, shrimps, and crabs, among many others. Disturbed only by occasional storms, the corals here blanketed all available rock surfaces—walls, pinnacles, the knobby protuberances known as bommies, deep channels, shallows—each providing shelter and food for invertebrates and fish. Fields of branching, table, and staghorn corals were broken up by an incredible diversity of smaller formations of other corals.

Parts of the reef were covered with bright pink patches of crustose coralline algae. These algae depend on fish to keep competing algae from growing over them. (Jim Stringer)

Soft corals of the genus *Sarcophyton*. (Randi Rotjan)

Nowhere was this profusion more dramatic than on the leeward reef of Kanton atoll. Here, the combination of strong tidal flow, low waves, and bright sunlight created the perfect conditions under which corals could flourish, forming banks of table corals that spilled down the reef slopes.

With each ebb and flow of the tide, the channels between the lagoons and the open ocean revealed a different exuberant cross-section of marine life. The channels heaved with parrotfish and snappers feeding in the changing currents. The exchange of cool, clear oceanic waters and warm lagoon waters rich with plankton and fine sediment resulted in plankton blooms in the channels, and plankton feeders such as manta rays swarmed through the current at the channel mouths, feeding on the concentrated blooms of food swept back and forth by the tides.

Inside the channels and away from the swirling tidal waters, the serene lagoons remain sheltered from the constant tidal influx. From the extensive lagoons of the large atolls to the central depressions on the small islands formed by the dissolution of carbonate rock by rainfall, the lagoons provide windows into the long geological life of an atoll. As the small islands sink below the surface of the sea—as first explained by Charles Darwin—the depressions become flushed by the tides, which open up space for corals to grow. Over time the lagoons expand and deepen to the point that they can host everything from perfect coral fields in the well-flushed margins of the channels to turbid, murky corners in the far reaches distant from the channel mouths.

A manta ray feeding on plankton. Where tides and currents concentrate plankton in dense quantities, rays can gather in large numbers to feed. (Cat Holloway)

The dense and varied coral growth sets the stage for equally dense and even more varied marine life. Coral reef fish, such as red anthias (subfamily Anthiinae) and banded dominofish (family Pomacentridae), hover in thick clouds around branching coral heads. Carpets of anemones host a multitude of anemonefish. Hawkfish stand guard on the coral heads, ready to challenge any diver's approach, while butterflyfish dart singly or in large groups over the coral meadows.

During all the Phoenix Islands expeditions, we observed abundant fish life on the sheltered reefs: shoals of roaming predators such as snappers, emperors, and trevallies, as well as herbivores such as parrotfish and surgeonfish, which graze by scraping algae from the coral. Larger predators—the sharks, dogfish, tuna, and schools of barracuda—also patrolled these reefs, in greater numbers than we had seen anywhere else.

On the steeper slopes of large rock surfaces, such as those we found at Rawaki and Birnie, the coral community showed a diverse mix of species with an astonishing range of growth forms—branching, massive, tabular, plating, encrusting,

Corals provide habitat for a wide range of reef fish (such as this hawkfish) as well as invertebrates. (Jim Stringer)

Butterflyfish grazing on the polyps of a branching coral. (Cat Holloway)

and solitary mushrooms. As we swam over the reef, we wondered, why all these growth forms? There is no clear answer, except that each growth form does particularly well under different conditions: branches grow fastest and make the most of abundant light in the shallows, massives form a robust skeleton resistant to breakage, plates grow outward from deep vertical walls to catch the most light, mushrooms grow loose on the bottom and do well under frequent disturbance because they can right themselves after being turned upside down. But these forms can all grow together, too, forming a reef with diverse structures.

Compared with the tranquil abundance and profusion of the sheltered western reefs, the communities on the exposed southern, eastern, and northeastern faces appear much more chaotic. On the southern sides of Nikumaroro and Enderbury, 10- to 12-foot (3–4 m) waves pound the shallow reef protecting the island. During our dives each wave pulled us up and down the near-vertical wall, even when we dove to depths beyond 100 feet (30 m) to escape their pull. From diving other reefs around the world, I knew that the shallowest parts would be composed of fields of broken, fast-growing corals. These corals grow during calm periods, then get destroyed by rough waves during weather cycles that bring storms to the islands. The cycle is repeated again and again, resulting in patches of vibrant new coral growth between rubble fields of the old.

Looking downslope, I could see the chutes through which the coral rubble—from small branches to rocks weighing hundreds of kilograms—spills down the steep slopes, catching occasionally as it makes its inevitable way downward into deeper waters. On the journey downward, which may take minutes, months, or even years, the rubble passes through recovering communities of hard and soft corals, sea fans, and algae, representatives of the biological and geological life of the atoll.

Compared with other remote island groups in the area, such as the Gilbert Islands to the west and the Line Islands to the east, we found a high diversity of both coral and fish species in the Phoenix Islands. Kanton proved to have the largest area of coral reef, with the highest number of species both within the Phoenix group and of the other islands surveyed so far.

The reefs of the Phoenix Islands gave me my first chance to see a nearly untouched atoll reef system. Here, over two hundred species of coral supported over five hundred species of fish. Of the corals themselves, two species represented range extensions—known species in a new location—and others were rare species known only from places where they were under threat. To see the reefs of the Phoenix Islands was to recognize instantly that they must be sheltered from the damage the modern world inflicts on nature.

Having seen this amazing pristine habitat, I was energized to work with Greg Stone and others to protect the coral reefs of the Phoenix Islands. With Stone and the owners of Nai'a Cruises, we began planning an expedition to gather more data and make a case for establishing a new marine protected area around the islands. Well aware of the growing threat of warming climates around the world, we nevertheless mistakenly thought that such a remote and pristine location might be a refuge from changing climates elsewhere. None of us suspected the changes the next two years would bring. Under the surface, even the waters of the Phoenix Islands were warming.

David Obura, Sangeeta Mangubhai, and Paul Nielson measuring plate corals. Measurements provide an estimate of coral size (a proxy for age) and condition. (Paul Nicklen)

Obura and Mangubhai entering data on corals aboard *Nai'a*. Measurements of coral diversity, size, and abundance are important to put PIPA in a comparative context. These data become part of the larger body of science about biodiversity across the Pacific. (Cat Holloway).

Corals and Climate Change RANDI D. ROTJAN

Corals worldwide are under siege. Reefs are being damaged by trawling, over-fishing, and other destructive practices, but such damage pales in comparison to the profound changes that are already being wrought by a different menace: climate change.

Given their distance from human settlements, the reefs of the Phoenix Islands should be less vulnerable to the threat of climate change than reefs elsewhere in the world. They should exhibit less change, or be changing less rapidly. At least, that's the theory. Yet the worldwide reach of human impact extends even to this remote corner of the globe.

In 2002 and 2003, while plans were under way to establish the Phoenix Islands Protected Area, high ocean temperatures in the Phoenix Islands caused a catastrophic coral-bleaching event, resulting in a steep decline in coral bio-

SIDEBAR 6.1: Corals and Our Warming Planet

As our planet warms, its increasing temperatures threaten coral reefs not only by warming the ocean, but by raising the sea level.

Rising temperatures raise the level of the ocean in two ways. First, as temperatures rise, land-based ice near the poles and on mountains melts, flowing into the ocean and adding to its volume. Second, seawater expands as it warms—a phenomenon called thermal expansion. Computer models suggest that sea levels may rise as much as 2 feet (60 cm) over the next century, or 2 inches (6 cm) in the next decade. If reefs are submerged below a critical depth, the depth of the water column will deprive them of the sunlight they need to live. Slow-growing coral species may not be able to "keep up" with rising sea levels.

Rising temperatures also affect reefs through the phenomenon known as coral bleaching. During periods of high ocean temperatures, such as those that occurred across the world's oceans between 1998 and 2010, corals experience episodes of mass bleaching.

Corals get their brilliant colors from zooxanthellae, simple plantlike cells that form symbiotic relationships with individual coral polyps. The zooxanthellae make food by photosynthesis, which they share with the corals. This algae-animal relationship is what makes coral growth in nutrient-poor waters possible. When corals turn white, the zooxanthellae either have lost their ability to photosynthesize or have abandoned the coral animal. Either way, bleaching is a stressful state that tropical corals can handle for only a short time. Without their symbionts to make food for them, the corals begin to starve, and if the zooxanthellae don't return quickly, the corals die.

Once the coral tissue is dead, the hard coral skeleton is bare. Nature abhors a vacuum, and the dead corals are soon covered in filamentous algae. Depending on the ecological state of the reef and how disturbed it is, this starts a long process of succession and recovery. The best scenario is a return to the original coral-dominated state via settlement of coralline algae, which give new corals a chance to grow. But other scenarios exist in which a degraded reef community takes over due to poor water quality or a fish community depleted by overfishing, resulting in dominance by turf or fleshy algae or other filter-feeding invertebrates such as sponges or soft corals.

Bleached corals. (Cat Holloway)

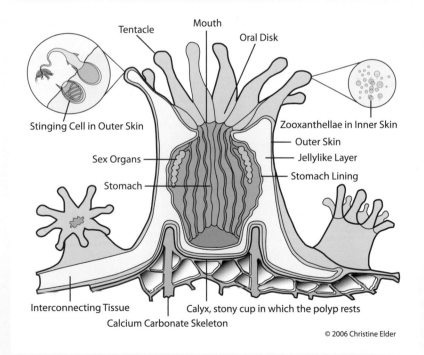

Anatomy of a coral polyp. (Christine Elder)

diversity. In some places more than half the corals were lost. Bleaching in corals can be caused by many stressors—diseases and changes in salinity, sunlight, and water temperature—but whatever the cause, the result is the same. Corals turn white, an unmistakable sign of stress.

The bleaching event in 2002 and 2003 was one of the most severe ever recorded. For seven months, temperatures remained above the threshold considered tolerable for a stable coral-algal symbiosis: a sea surface temperature of 86°F (30°C) lasting a week or longer. Throughout the Phoenix Islands, corals turned white, and many died.

Unlike bleaching events elsewhere in the world, this one was unique because it was not exacerbated by local human impacts. Think of a reef as a patient who first gets the flu and then gets pneumonia. In the case of the Phoenix Islands, the reefs weren't subject to any such "complications" caused by the stress of local human actions. Scientists observing the aftermath of this severe bleaching event were eager to see how it would unfold in the islands. With no people around to exacerbate the disaster, would these reefs recover more quickly than reefs elsewhere? I was fortunate to be one of the scientists chosen to find out.

Nikumaroro, September 2009

Before I ever dove into the waters of Nikumaroro, I had heard a lot about the Phoenix Islands from my colleagues at the New England Aquarium. Given what I knew about the 2002–3 bleaching event, I approached my first dive in the Phoenix Islands with caution. How would the reefs look six years after the disastrous bleaching event? Excited as I was to finally see these reefs for myself, part of me almost didn't want to enter the water. Would I find a healthy, thriving, recovering reef? Or would I see a coral graveyard?

I jumped in and saw some rubble reefs—broken coral skeletons with very few live, healthy corals. My heart sank. Here, in the most remote part of the world, were coral ghost towns. As I swam around, taking samples and recording data, I found myself struggling to hold on to my optimism about the fate of the ocean. Ever the scientist, I thought, "Well, at least we'll have a geological record of global change forever preserved in coral skeletons." But if I was going to find a coral graveyard on every dive, my first expedition to the Phoenix Islands was going to be a long, depressing trip.

Shaken, I tried to hold on to some hope. After all, any coral reef scientist knows that there's huge variation from site to site. I knew that these reefs had experienced incredibly patchy mortality—as few as 20 percent of species in some spots, 100 percent in others. Surely not every reef would be rubble?

The next day came and, still full of dread, I hopped into the water. To my delight and relief, I saw stunning reefs covered in live corals, teeming with fish, patrolled by turtles and manta rays and sharks. The difference was astonishing. Mingled with my relief was the dawning realization that these resilient reefs offered us the opportunity to learn something new and exciting: why reefs recover at different rates in the *absence* of intense local human impact.

Hopeful signs of a coral comeback: in 2009 juvenile corals were observed recolonizing the reefs bleached in 2002–3. (Randi Rotjan)

Shifting Baselines

Randi Rotjan inspecting and measuring corals. (Jim Stringer)

Even though I was on the 2009 expedition as a dispassionate scientist, I was unable to suppress my wonder and amazement and joy at seeing such beautiful, and apparently fully recovered, reefs. When I regained my scientific composure, I asked David the obvious question: "So, how do these reefs compare to their former, pre-bleached selves?" I was shocked at his answer. Qualitatively, David thought that they were only halfway restored to their former glory. *Only halfway?!* How was it possible that such a beautiful reef could be twice as beautiful? I was stunned—until I realized that I had almost been a victim of shifting baselines.

As each new generation experiences the ocean, we form our view of what is healthy based on what we've seen with our own eyes. Thus, when we think about conserving the ocean, we seek to maintain or restore it to that baseline state, our idea of what a healthy ocean looks like.

SIDEBAR 6.2: A Living Laboratory

Dr. Gregory S. Stone in the Phoenix Islands in 2002. (Cat Holloway)

The next decade may see the remote Phoenix Islands, once home to guano miners and a NASA tracking station, turned into a living laboratory where scientists can come to gather data about the ways ecosystems respond to climate change.

Climate change is already here, bringing profound changes to the ecosystems on which billions of people rely. Scientists need to know how healthy ecosystems respond to changes such as longer or more intense El Niño events. These events are part of a climate cycle known as the El Niño Southern Oscillation, or ENSO. El Niño events bring unusually warm temperatures to the waters along the equator. The reefs of the Phoenix Islands experienced an extreme El Niño event in 2002–3, the worst so far recorded, which killed over 60 percent of the corals. Though tragic, this event provided scientists the opportunity to measure reef recovery. Most reefs suffering from catastrophic bleaching show few signs of major recovery a decade later, but thus far, the reefs of the Phoenix Islands appear to possess remarkable potential to recover. As of 2009 they appeared well on their way to recovery from the catastrophic warming event.

These islands may have much to teach us about marine ecosystems' ability to rebound from the stress of warming ocean waters. Because they are so remote, the Phoenix Islands are one of the few places where the effects of El Niño events can be observed apart from local-scale human stressors such as overfishing, pollution, or human-induced disease. When compared with similar island groups in the region, such as the Line Islands or the Gilbert Islands, the Phoenix Islands provide a powerful natural laboratory, one of the only places on Earth where global and local events can be decoupled.

But these islands' recovering reefs are only part of the story. Beginning in 2012 scientists from the New England Aquarium, the Woods Hole Oceanographic Institution, Scripps Institution of Oceanography, and Conservation International will come to the Phoenix Islands to take advantage of this powerful natural experiment in a decade-long undertaking called the Phoenix Islands Initiative. Working with the government of Kiribati, researchers will compare the Phoenix Islands with its sister groups, the neighboring Gilbert and Line Islands, to see how human population and development affect the ability of reefs from recover from events like El Niño. They will also compare the Phoenix Islands with the northwestern Hawaiian Islands and other archipelagoes, assembling a detailed picture of ocean health throughout the Pacific.

Using sonar and remotely operated vehicles, these teams will explore the reefs and beyond, mapping the island slopes, deep ocean floor, and seamounts and conducting blue-water surveys of the open ocean. They will tag large oceangoing fish and take DNA samples to better understand connectivity—where species arise, how they spread, and how they eventually move about the globe.

Other researchers will use the Phoenix Islands and its neighbors to study the region's ecological economics: the ways in which nature and the humanized landscape interact as a coupled system. Ecological economists will compile data on the habitats that are shared by people and other species and on the ebb and flow of food, materials, and employment. The uninhabited Phoenix Islands will provide an essential control. A better understanding of the role of people in the ecosystem over time will help economists develop models to help policymakers make decisions that will preserve the islands of the Pacific for the peoples who depend on them, now and for generations to come.

From both the ecological and economic perspectives, the Phoenix Islands are the critical link to understanding how habitats respond to global change without human stressors. As other reefs worldwide struggle for survival, they are plagued with simultaneous local and global stressors. Scientists, economists, politicians, and ordinary citizens want the answers to tough questions: If we remove given stressors (such as sedimentation, overfishing, or pollution), will reefs recover? These important questions are unanswerable in places that are suffering from a deluge of stressors. The Phoenix Islands provide a blueprint for global reef recovery because they stand alone as a chain of islands barraged by intense climate events (El Niño), but unencumbered by daily human activity. If the Phoenix Islands can recover from the worst coral-bleaching event ever recorded in human history, it provides hope and insight for reef recovery elsewhere.

We don't yet have the science and technology to restore damaged and degraded reefs. But by studying the Phoenix Islands and their ability to rebound from bleaching events and other forms of climate change, scientists hope to develop the data and tools to allow us to better protect and manage imperiled ecosystems everywhere.

But what if our baseline is skewed? What if "healthy" to us isn't "healthy" in an absolute sense because we have never seen a healthy ocean with our own eyes? How do we recalibrate our baseline to the true baseline? How do we even determine the true baseline?

Herein lay my problem: without my colleague David to help me recalibrate my idea of a healthy Phoenix Islands reef, I might have been fooled into thinking that these reefs had fully recovered from the 2002 bleaching event and currently stand at the peak of their glory. But because of the careful observations made by David and the other scientists who had come to these islands before me, my view of this reef is more tempered. Based on the limited observations from the 2009 expedition, I can now cautiously state that these reefs are on their way to recovery, but that recovery is not yet complete. Will they one day be restored to their former splendor of species diversity? Only time and data will tell.

A Reef Rises from the Ashes

The bleaching event in the Phoenix Islands offered coral biologists around the globe a real-time natural experiment in reef recovery. The eight islands vary considerably in size and configuration, in the presence or absence of lagoons, and in their exposure to wind and waves. Not all the reefs responded to the warm spell in the same way, and they recovered at different rates and in different manners. We were eager to know what factors affected the rate of recovery. Which corals led the recovery? What differences were there between islands? What roles, positive or negative, did island size and the presence of lagoons or shipwrecks play? How did herbivores such as parrotfish function to prepare the coral rubble for recolonization? The insights we gain from monitoring the complex dynamics of coral reef recovery will move the science of reef conservation forward.

The State of the Reefs

The Phoenix Islands reefs are showing some promising signs of recovery and re-growth. In 2009 we observed lots of young corals as well as an extremely healthy fish population, with large schools of each of the major ecosystem players. The fish are doing an effective job of keeping the algae from overgrowing the corals, and so we see a reef primed for recovery. There is a lot of available space for corals to grow on, there is lots of crustose coralline algae for young corals to settle on, and it seems that there are enough remaining adult corals of most of the pre-2004 coral species to reseed the population. Part of the excitement and challenge of the 2009 expedition was to debate and discuss our observations in real time and to try to reconcile our qualitative observations with the quantitative data gathered on every dive. Overwhelmingly, our story is one of optimism and hope, one of the rare "happy" coral reef stories in science at the moment.

There has been much discussion of how to address the dire state of reefs around the world. Suggestions put forward include more and better measures to protect reefs, legislation to stop overfishing, and the creation of marine reserves to give marine wildlife a buffer against the effects of climate change. Despite all of the scientific evidence for climate change, popular acceptance of the scientific reality is far from universal, and the message of a climate emergency in the ocean has not been received globally. Most countries around the world do not do enough to protect their oceans. Ocean stewardship is still a relatively new concept for some, despite the encouraging recent creation of some large marine reserves in Hawaii in 2006, the PIPA itself, and the Chagos Islands marine reserve in 2010. With hard work over the next decades, it may still be possible to achieve our goal of healthy reefs worldwide. There are compelling hints of hope. As of 2010, however, the majority of the world's coral reefs are struggling for their very existence and facing the looming threat of ocean acidification.

Over the next decade, David and I hope to return to the Phoenix Islands with other coral scientists to move our investigations from simply describing what species are there to understanding how the reef community works at the process level. We will move our research questions from the local scale to the global scale, hoping to piece together how corals evolved and spread throughout the Pacific by analyzing their DNA.

Coral biologists working throughout the Pacific hope to learn how different coral species and genetic lineages differ in their responses to ocean warming. To piece together the story of coral resilience, they will take coral cores to see how reefs responded to warming events in the distant geological past. They are also studying fluorescent corals to see whether the ability to fluoresce confers some immunity to bleaching events, as suggested by work at Lord Howe Island in Australia.

Plans for future work within the PIPA are still taking shape, but the priority will be to place the reefs of the Phoenix Islands in their larger regional and global context and to share their lessons with the scientists and conservationists working to save reefs around the world.

Expedition Diary: A Roller-Coaster Ride on the Reef's Edge, Manra, 2000   DAVID OBURA

On our first expedition in 2000, we explored the coral communities along the exposed northern shore at Manra. After suiting up in diving gear, I ventured into the water, joined by fellow coral biologist Sangeeta Mangubhai of World Wildlife Fund's South Pacific Program.

The northern and northeastern slopes of the Phoenix Islands are less subject to the mercies of massive waves than the southern sides, but are still shaped by strong currents. On Manra, Sangeeta and I took a roller-coaster ride along the reef edge as we found ourselves carried along by one of those currents, moving from shoals of darting and feeding jacks to fast-swimming and agitated sharks and then shoal after shoal of fusiliers. We sped past curious red bass and streams of snappers, startled turtles as they hurtled by on the current, and swept past dolphins feeding on the abundant fish schools. As the current pulled us along, the dolphins disappeared into the blue, replaced by a massive shoal of barracuda that darkened the light as we approached them, then kindly hollowed into a doughnut shape to allow us through. Our roller-coaster ride ended dramati-

SIDEBAR 6.3: Ocean Acidification

Dr. Christopher Sabine, from NOAA's Pacific Marine Environmental Laboratory, standing next to a CTD (conductivity, temperature, depth) rosette, an electronic instrument used by oceanographers to measure ocean acidification. (NOAA)

The ocean acts like a giant sponge, soaking up excess carbon dioxide from the atmosphere. But lately, with 7 billion people and 600 million vehicles producing carbon dioxide, humans are producing more carbon dioxide than the ocean can soak up. It's estimated that the ocean has become 30 percent more acidic since 1760, the beginning of the Industrial Revolution. Experts fear that the rate of acidification will accelerate in the coming decades.

The shells of many marine creatures are made of calcium carbonate, and many of these marine calcifiers—including crustaceans, mollusks, corals, and some plankton—are vulnerable to even small changes in the ocean's pH. With large changes in pH, they may lose their ability to form shells. Therefore, increasing acidification is likely to cause a cascade of effects in marine food webs, with dire consequences for food security in regions that depend on seafood to feed their populations as well as for major economic sectors based on ocean resources. Increasing ocean acidification could lead to an environmental and humanitarian crisis of unprecedented proportions.

Some of the most sensitive species are the reef-building corals, whose very existence is based on efficient calcification. As reefs corrode, populations of fish and other high-protein food sources will decline, and low-lying countries like Kiribati will experience erosion and inundation, no longer protected by the buffer of a coral reef.

Ocean acidification and global warming are two separate consequences of increasing carbon dioxide levels in the atmosphere, but these two forms of human-generated climate change are interacting to cause unprecedented changes to the planet.

cally when a rip current off the island shot us out into the Pacific through more shoals of fusiliers, a few more inquisitive sharks, and up to the surface under the ever-watchful eyes of the skiff crew.

A few days later, when Sangeeta and I dove the reefs on the islands with lagoons, we saw communities with entirely different compositions of organisms. In the mouth of the channel at Orona, we could clearly see evidence of reef life cycles—changes with the tides, changes with the phases of the moon, changes over the time scale of decades and centuries. On our first trip through the channel, in the ankle-deep channel mouth, we saw nothing when we motored in and out of the lagoon at mid-tides. But one day, just before the end of the ebbing tide, as Sangeeta and I were finishing a dive on the outer reef, we decided to take an exploratory snorkel into the channel mouth. We pulled ourselves forward hand over hand in water barreling out through the channel, barely deep enough to lie in. As I poked my head above the fast-flowing current to get my bearings, I was dazzled by what I saw in front of me: more than a hundred small black triangles poking above the surface of the water. Some were languidly undulating in the water; others were darting back and forth. With more surprise than fear, I put my head back under the water just in time to see a shape come hurtling past me in the current. It was the pale brown-and-white-patterned side of a small blacktip shark, probably more startled by me than I was by it.

A school of juvenile blacktip sharks had congregated in the channel at the end of the ebbing tide to prey on the surgeonfish and parrotfish swimming out of the lagoon on the tide after feeding in the shallow water. With little depth of water for escape, the fish were easy prey for the sharks. The whole team came back later to watch and photograph this short and exciting event.

7 ECOTOURISM IN PARADISE KANTON

ROBERT BARREL

Sir David Attenborough, speaking of the Galápagos Islands, called ecotourism "a necessary evil," but for many threatened ecosystems, such managed tourism may provide the only practical alternative to more destructive forms of development for emerging economies in dire need of income.

Ecotourism in the fragile and remote Phoenix Islands may be the most dependable source of income available to protect and conserve the islands, reefs, and myriad sea creatures there. It was ecotourism that first brought the Phoenix Islands to the world's attention. Without tourist divers or citizen scientists to pay for the initial exploration voyages, scientists and conservationists would not have had access to this extremely remote archipelago.

A dancer being outfitted in a traditional costume before a performance on Orona in 2002. (Cat Holloway)

Nai'a's first exploratory expedition to the Phoenix Islands was in 2000, but I realized the value of ecotourism as a conservation tool shortly after Nai'a Cruises began running our 120-foot liveaboard dive boat in Fiji. Some of the lessons learned there may be applicable to low-impact, high-value tourism in the Phoenix Islands.

I grew up in Hawaii and discovered early on that the natural beauty of that place could best be appreciated from well offshore, where the air was clean and the high-rises and tourist rental cars vanished from view. In close to the beaches and waterfalls, the fragrance of plumeria gave way to exhaust fumes, and the trails were littered with junk-food packaging. The best way to appreciate Hawaii was to learn to sail and escape to the less popular, less populated places.

I sailed small boats around the most isolated Pacific islands for eleven years before attending the University of California, Santa Cruz (UCSC). During those years at sea, I favored out-of-the-way ports: If there was another boat at anchor when my companions and I came in through the pass, we were disappointed. If there were several, we were likely to turn around and keep on sailing. One of the countries we most enjoyed cruising in was Fiji, with nearly four hundred islands to explore, very few cruising yachts, and a population of well under a million. So it was no coincidence that my wife and partner, Cat Holloway, and I chose Fiji as our home port when Nai'a Cruises began operations in 1993.

Our original goal for *Nai'a* was to combine a tourist diving business, as a means to pay for the ship, with the long-term scientific exploration of dolphin culture, the subject I had studied for my anthropology degree at UCSC. Unfortunately, Fiji's resident pods of dolphins chose to live where the diving was lousy, so I couldn't spend enough time among them to develop the scientific data I needed.

We set our sights instead on the diving. Fiji is one of the world's great destinations for exploratory diving. It didn't take long before Cat and I started expanding our route and scouting new dive destinations, motivated as much by our passengers' excitement as by our own desire for discovery. As we began to develop a repertoire of magnificent dive sites, we also began to understand what combination of natural features and functions made a dive site memorable and rewarding. We learned, for instance, that fish tend to congregate where the water flow over and past the reef is the strongest and in channels or near spurs of reef that protrude into the current; that reef topographic features that are attractive to divers, such as archways, aren't necessarily appealing to fish; and, most importantly, that healthy fish populations start with healthy coral reefs. You won't have one without the other.

A few academics did take us up on the chance to use *Nai'a* as a base for research on Fiji's reefs. Peter Newell, senior biologist at the University of the South Pacific (USP), had us peering into the titillating world of nudibranch and flatworm sex. (Did you know that a nudibranch has a penis located where you might expect to find its right ear?) Norman Quinn initiated a fascinating study into ocean currents and water temperatures. Since 1997 small temperature recorders set throughout central Fiji and monitored by *Nai'a* crew and passengers have provided calibration for satellite-derived measurements of sea surface temperatures. The temperature recorders have also helped to determine

Nai'a at anchor.
(Mark Snyder)

whether cold, nutrition-rich deepwater upwellings might explain central Fiji's unusually robust and diverse reefs. The answer so far is probably not. Something else is going on. What that something else might be awaits further investigation, in which the citizen scientists aboard *Nai'a* may well play a part.

Stan Flavel, a master's student at USP, had us collecting sponges, specifically *Jaspis coriacea*. This simple orange sponge contains a miraculous antitumor agent dubbed a "beqamide," after Beqa Lagoon, where the highest yield of the chemical had been identified. We were able to recruit the keen eyes and intelligent minds of our passengers to record crucial information about the sponges and their natural habitat. Each sample was photographed, and the site growth and regrowth (after sampling) was monitored carefully. The time, day, depth, and place; water clarity, salinity, and content; and current and light levels were recorded. In addition, each sponge leads an active private life that Flavel needed to understand. Who eats it and when? Who are its neighbors, friends, and enemies? How does it earn a living? How does it look first thing in the morning? Who does it "date" at night?

Divers as Citizen Scientists

The more we engaged our passengers in scientific projects aboard *Nai'a*, the more interested they became in diving in general. Liveaboard divers are, as a group, particularly well suited to act as "citizen scientists." They choose to spend four or five hours a day scuba diving on reefs up to half a world away from their homes, so a profound interest in the underwater world is a prerequisite. A week or ten days on a liveaboard in the middle of the Pacific doesn't come cheap, so *Nai'a* divers are often successful, goal-oriented people. Often they own and know how to use the very latest video and still photography equipment. We discovered that we could give these affluent, motivated, and extraordinarily capable people an intellectual challenge, one that allowed them to make a contribution to science and ocean conservation. It was a classic win-win situation.

This symbiotic partnership between Nai'a Cruises, our diving passengers, and research scientists continued as the scientific community learned what tourist divers had known for some time: that some of Fiji's reefs are unexpectedly diverse and healthy. During our first four years in Fiji, we explored the whole archipelago, but in 1997 our focus began to shift from exploration to protection. The World Wildlife Fund, the Wildlife Conservation Society, and the Coral Reef Alliance all took an interest in Fiji's marine resources, an interest that we fostered by running research and donor trips for them. That collaboration led to the establishment in 1997 of the Namena Marine Reserve, a 27-square-mile (70 km²) reserve around the island of Namenalala. The Namena Marine Reserve is part of the Vatu-i-Ra Seascape, Fiji's first science-based network of marine protected areas. Workshops on community-based ecotourism business design are giving the local community members the training they need to launch successful ecotourism enterprises.

The Vatu-i-Ra Seascape project has increased awareness and appreciation among governments and conservation organizations for the unique and often robust coastal environments in the central Pacific, an area often overlooked in the competition for research funding. This experience was to stand us in good stead when we began to bring *Nai'a* to the Phoenix Islands.

Citizen science at work: A *Nai'a* passenger surveying near a formation of giant *Porites* corals. (Cat Holloway)

A passenger from *Nai'a* serving kava during a visit to a Fijian village. Such visits in Kiribati might help fund conservation efforts within the PIPA. (Cat Holloway)

It was through Peter Newell that we met Greg Stone, a marine scientist then working for the New England Aquarium. In 1998 we invited Greg and his wife, Austen, to join us in Tonga for one of our first passenger-funded research cruises among humpback whales. Greg recognized right away the benefit of a well-equipped ship whose basic operating costs are covered by paying passengers. During the evenings in Tonga, while recovering from a long day of snorkeling with whales, we began to brainstorm projects on which we could collaborate. But the one that got all of us so excited that we could barely sleep was a return to Nikumaroro and a first-time exploration of the rest of the Phoenix Islands. As luck would have it, Cat and I had been contemplating such an expedition only a month before.

Searching for Amelia, Stumbling on Paradise

We "discovered" the Phoenix Islands for the first time in 1997 when TIGHAR (The International Group for Historic Aircraft Recovery) chartered us to support their search for the missing aviator Amelia Earhart on Nikumaroro. The expedition was entirely focused on the search for the remains of Amelia's airplane on the island and in the lagoon. Scuba diving (or even swimming) in the open ocean was strictly prohibited by TIGHAR's management because of the robust shark population. But there is no channel into the lagoon at Nikumaroro, so, without a place to anchor on the outside of the atoll, *Nai'a* was forced to remain under way, burning precious fuel day and night. It was in everybody's best interest that we find a place to moor the boat. TIGHAR director Ric Gillespie reluctantly gave us permission to dive in order to set a mooring in the only place where protection was offered by the steep reefs close to shore.

When we dropped into the water, we were immediately surrounded by a vast school of jacks and about forty curious gray reef sharks. It was intimidating and exciting. By the time we made our way to the bottom, the sharks' curiosity had abated, but now we were distracted by maybe a dozen mantas tumbling in the current as they fed in the nutrient-rich water flushed out of the lagoon by the huge waves. On a subsequent dive, we had to push curious turtles out of the way to actually set the mooring. And for the rest of *Nai'a*'s time at Nikumaroro, Cat and I convinced Gillespie that, for the safety of all concerned, we really needed to dive on the mooring at least twice daily to make sure that it was holding!

While TIGHAR was looking for Amelia, we seemed to have stumbled on a diving paradise. The amazing abundance and incredible beauty of the reef at Nikumaroro was still fresh in our minds when our good friend and frequent passenger Kandy Kendall asked us, during one of our evenings under the stars back in Fiji, where she could go that was truly unexplored. Cat and I looked at one another, looked back at Kendall, and said together, "Nikumaroro."

We took the charts inside and laid them out on the big table in the main salon. Lying almost exactly 1,000 miles (1,609 km) NNE of Fiji, Nikumaroro is remote from everything, with the nearest airport located 700 miles (1,127 km) away in Apia, Samoa. Even the ancient Polynesian navigators didn't stop in the Phoenix Islands for long because they are not on the way to anywhere and there is no reliable source of drinking water there. There weren't even coconuts on the islands until early colonists planted them. The Phoenix Islands are probably the least explored islands in tropical or temperate waters.

Kendall was instantly committed, and she offered right then and there to pay for half the cost of an expedition, provided we could put together a team of reputable researchers who could make the voyage to the Phoenix Islands scientifically valuable.

Fast-forward a month, and there we were in Tonga with Greg and Austen Stone. Suddenly we had the bones of our research team for a Phoenix Islands expedition. Greg and Cat both went home from Tonga and got to work on planning the expedition, Greg on identifying a solid scientific team while Cat set about recruiting citizen scientists to pay the remaining cost of the expedition. The problem soon became limiting the number of people we could take because there wasn't enough room for everyone who was interested.

One of the citizen scientists who committed to the Phoenix Islands expedition very early on was Bruce Thayer. While staying with us in Fiji a few months before the expedition, Thayer recognized that we had good scientists and good divers on the team, but we didn't really have a strong conservation person. We called the head of WWF South Pacific, and Thayer issued a financial challenge: if WWF would be willing to send a representative, he would pay half the cost and Nai'a Cruises would pay the other half. WWF agreed to send Marine Conservation Officer Sangeeta Mangubhai, who brought to the expedition a passion for fieldwork and a diplomatic, professional approach to conservation management.

So the first ocean science expedition to the Phoenix Islands was born. An ecotourism venture funded entirely by citizen scientists, the 2000 expedition proved so successful that Greg Stone was able to arrange full corporate funding for a longer expedition two years later.

Responsible Ecotourism in the Phoenix Islands

Now that the Phoenix Islands Protected Area has been established and declared a World Heritage site, there is a clear opportunity for the government of Kiribati to develop a sustainable ecotourism operation in the islands. How can the Phoenix Islands be responsibly opened up to modest ecotourism ventures?

The United Nations Economic and Social Commission for Asia and the Pacific (ESCAP) issued a report in 2003 in which it outlined the steps that needed to be taken to make ecotourism successful in Kiribati. Tourism officials in the Kiribati government have already identified possible ecotourism activities that might be developed elsewhere in Kiribati, including catch-and-release game fishing for species such as bonefish, bird-watching, tours of traditional villages, and tours of World War II–era sites such as the battlefield on Tarawa. There are plans for eco-lodges built from local products by local labor and for tourists to stay with local families to experience traditional I-Kiribati life. Day trips for tourists based on cruise ships are also being considered. All of the ventures proposed so far would take place outside the Phoenix Islands Protected Area.

Some of these ecotourism activities would suit the Phoenix Islands, but they will never see the level of ecotourism developed, for instance, in the Galápagos Islands. The Phoenix Islands are just too remote. But it is their remoteness that makes them attractive to a certain type of tourist, the same type of tourist— not coincidentally—who has supported ocean ecotourism in Fiji and Tonga for nearly two decades.

SIDEBAR 7.1: Midway Atoll: A Model for Tourism in the PIPA?

The northwestern Hawaiian Islands are home to millions of nesting albatross and other seabirds as well as to endangered Hawaiian monk seals, green turtles, and spinner dolphins. They also boast some of the most extensive and healthy coral reefs in United States waters. In 2006 this chain of islands, reefs, and submerged atolls was declared the Papahānaumokuākea Marine National Monument (PMNM) and closed to most recreational and commercial activities.

In 2008 the Oceanic Society began running eight-day ecotours to Midway Atoll National Wildlife Refuge, which lies within the PMNM. The Oceanic Society limits its tours to sixteen participants throughout Midway Atoll.

Access is by turboprop from Hawaii, a 4.5-hour flight. Visitors stay in renovated World War II barracks; accommodations are basic. On arrival, everyone must complete a National Wildlife Refuge orientation to learn how to observe nesting birds and seals without disturbing them and how they can contribute to island stewardship through service projects. In the company of naturalists, tour members explore historical sites from World War II, observe nesting

SIDEBAR 7.1 (*continued*)

birds and seals, and snorkel in the lagoons. Each visitor is provided with a bicycle to traverse the island with the least disruption either to the wildlife or to sensitive archaeological heritage sites. Tour members are encouraged to attend nightly natural history and conservation lectures as well as to participate in habitat restoration by removing invasive plants and collecting debris, such as fishing nets and plastic, that poses a hazard to nesting wildlife.

If ecotourism comes to the Phoenix Islands, it might well look like this.

Ecotourists on Midway Atoll in the Papahānaumokuākea Marine National Monument. (Wayne Sentman)

Ecotourism would depend on reliable access to Kanton by air. Improving the landing strip there might be a long-term option, but another solution might be the use of a modern twelve-passenger seaplane such as the Dornier Seastar. This form of transport would allow reliable visitor and staff transfer into and out of Kanton as well as small-scale ecotourism development of Orona and Nikumaroro. It would also increase the capability of I-Kiribati officials to monitor fisheries.

A low-impact lodge with safari-style "eco-tents" could be built on existing foundations at the Southside site of the former Pan Am hotel. Such a lodge would be designed for efficiency to minimize the reliance on outside supplies. The lodge would host a rotating group of scientists and other experts who, in addition to carrying out ongoing research for the PIPA, would interact with the guests in informal settings. Activities would include scuba diving, snorkeling in the lagoon, bird-watching, sailing on traditional I-Kiribati canoes, and perhaps catch-and-release bonefishing and big-game fishing.

The seaplane would allow small groups of divers or sightseers to make day trips to Nikumaroro or Orona, or even Enderbury and Manra in calm weather. Deepening and widening the existing tidal channels of the PIPA's outlying

Gavin Platz of Tie 'N' Fly Outfitters with Kiriata Tokiauea, local magistrate of Nonouti. Tokiauea spearheaded the change from destructive net fishing in the lagoon to traditional hook-and-line fishing. (Tie 'N' Fly Outfitters)

Fishing for bonefish (*Albula vulpes*) is a sport long popular in Hawaii, Florida, Texas, and the Bahamas. Bonefish don't taste very good, but they are considered one of the world's premiere game fishes by saltwater fly-fishing enthusiasts.

Bonefishing is the antithesis of charter-boat fishing for massive game fishes like marlin. Bonefish are relatively small, slender silver fish that congregate in shallow flats among sea grass beds, where they feed on small shrimps, crabs, and fish. To catch them, anglers need to wade into the shallows or balance standing up in a skiff, casting into the waters with a light line. Catching bonefish isn't easy. Millions of years of predation by such fish as barracuda have bred wariness into their DNA, and their silver coloring makes them very difficult to see in the shallows. Once they have taken the hook, they can lead even experienced anglers on a long chase before being landed; not for nothing does their Latin name mean "white fox." Therein lies the lure for dedicated saltwater fly-fishers.

Bonefishing is one of the oldest catch-and-release fisheries, one that attracts lots of well-heeled travelers to places they might not otherwise visit—places like Nonouti Atoll in the Gilbert Islands group of Kiribati.

In the language of Kiribati, "Nonouti" means "wake up early to go fishing." In 2010 the Kiribati National Tourism Office, in conjunction with a Brisbane-based saltwater fly-fishing enterprise, Tie 'N' Fly Outfitters, undertook four exploratory trips to see whether Nonouti was suitable for recreational bonefishing tourism similar to that already established on Kiritimati. The Australians toured potential fishing grounds, went bonefishing from a traditional canoe

SIDEBAR 7.2 (*continued*)

with the Nonouti magistrate, trained local people to serve as fishing guides, and donated supplies to the village schools and medical clinic. The tourism minister, Tarataake Teannaki, cautions that a fisheries conservation and management plan is needed in order to ensure that bonefishing tourism in Nonouti Atoll is sustainable.

If bonefishing in Nonouti proves successful, it's possible that Kanton could also be developed as a world-class bonefishing destination.

lagoonal islands would benefit both biodiversity and ecotourism, allowing boats to be left at the islands for the use of bird-watching or dive groups.

The other alternative for small-scale ecotourism in the Phoenix Islands would be entirely boat-based. One of the new breed of expedition cruise ships, able to carry fifty to a hundred passengers, could visit all or some of the Phoenix Islands with minimal impact. These vessels carry everything they need, from food and fuel to inflatable skiffs to ferry passengers ashore. Most of the same activities that a land-based eco-resort would offer could be offered aboard or based from the ship, including support for ongoing scientific research as well as fisheries enforcement.

Of the Phoenix Islands, only Kanton offers a safe all-weather anchorage for a ship, so it would make sense to base the ship there. Regular visits by a larger oceangoing ship could resupply Kanton's small caretaker population, or even a larger contingent if the ship were to be combined with a land-based lodge. Indeed, the airstrip at Kanton could be used to fly people back and forth from the ship, although a modern ship could make the passage to Samoa and back in about a day and a half.

The National Geographic Society partners with New York–based Lindblad Expeditions to offer small-scale "travel philanthropy," using expedition vessels to bring fewer than one hundred passengers at a time to such remote island destinations as the Galápagos and Ascension Islands. Experience from ventures such as these could instruct and inform the development of any eco-cruises to the Phoenix Islands.

While ship-based ecotourism would clearly have the least impact on the fragile islands of the Phoenix group, it would appear to offer fewer job opportunities for the I-Kiribati. Kiribati is beginning to send an increasing number of able sailors into the international marine fleets, however, and soon it might be possible to staff a ship for the Phoenix Islands almost exclusively with I-Kiribati.

The Phoenix Islands are unique from an ecotourism perspective: a new World Heritage site where, with the right infrastructure development, discerning and ecologically minded tourists could experience a pristine ocean environment that exists nowhere else. For those with the time and wherewithal to make it to the Phoenix Islands Protected Area, it would be a once-in-a-lifetime experience.

Utmost care would have to be taken in bringing even the most modest form of ecotourism to the islands, with conservation and biosecurity always taking priority, lest we love the Phoenix Islands and their natural treasures to death. Recent research has shown that the Phoenix reefs are remarkably resilient, but we should not and cannot take that resilience for granted.

Expedition Diary: Nikumaroro's Mysterious Allure, 1997

CAT HOLLOWAY

I first saw Nikumaroro in 1997, when Rob Barrel and I skippered a *Nai'a* cruise on behalf of a group trying to solve the mystery of vanished aviator Amelia Earhart.

The small, low atoll revealed its unmistakable palm tree crown from 5 nautical miles (9 km) away. At first we thought the smudge on the horizon was a cloud, or yet another mirage. But the charts and binoculars confirmed that we'd arrived at last at mysterious Nikumaroro—a remote island that doesn't give up its secrets easily.

An odd spiritual energy seems to encircle Nikumaroro like its nearly impenetrable reef. The island is profusely covered with *buka* trees (*Pisonia grandis*), traditionally thought to be sacred to the goddess who brought the gift of navigation

Nikumaroro's wooded coast seen from the air. (Ric Gillespie/TIGHAR)

to the Gilbert Islanders who settled here. Those Gilbertese settlers believed this island to be Nikumaroro, the legendary home of their most revered and potent ancestress, Nei Manganibuka. According to Kiribati custom, cutting the atoll's trees after dusk will incur the wrath of the spirits who dwell in them, as will the odor of any foreign trespasser. Set foot on Nikumaroro's glistening beaches, if you dare. But go no farther until you rub sand on your cheeks so that when the island spirits descend to investigate the intruder, you will smell of the island.

But Nikumaroro's mysterious aura doesn't stop with old legends. There are modern mysteries, too. During their five expeditions to Nikumaroro, *Nai'a* crew members have sworn they heard "ghosts breathing" while the ship sat silently moored offshore. Whales blowing, perhaps, but whales are rarely sighted there today. Visitors to Nikumaroro have also witnessed unexplained lights from the uninhabited shore at night. (There is no swamp gas on tropical atolls.) During our first trip there in 1997, both the crew aboard *Nai'a* and the shore party spread around the island reported hearing a twin-engine prop plane flying overhead. Yet not one of us actually saw it. The ghost of Amelia?

At least some of this superstitious fancy is treated seriously by scientists from The International Group for Historic Aircraft Recovery (TIGHAR). In their meticulous fifteen-year quest to prove that Earhart landed on Nikumaroro after becoming lost during the final leg of her round-the-world flight, TIGHAR researchers have endured Nikumaroro's relentless heat, ruthless storms, and vicious coconut crabs as they search for hard evidence to solve the Earhart mystery. They are drawn back to Nikumaroro at every opportunity, possibly more by its haunting charm than by the clues they have uncovered there.

Tom King, the chief archaeologist for TIGHAR's Earhart project, knows well the thrill of discovery as well as the poignancy of the island's unfulfilled promise. "Sometimes I worry that this island will never allow us to solve its mystery," King said. "We've learned so much here, but we understand so little."

The same could be said about the mystery that lies below the surface, in Nikumaroro's spectacular reefs. Because of the number of sharks in the water, TIGHAR's expedition leader, Ric Gillespie, forbade diving and snorkeling during our first encounter with Nikumaroro in 1997. But Nikumaroro's treacherous coast offered no safe anchorage for *Nai'a*, so Rob and I were granted just enough dive time to set a strong mooring for the boat. In those few blissful minutes underwater, we found ourselves rolled into a swirling mass of five hundred trevally and the probing rush of blacktip, whitetip, and silvertip reef sharks. We marveled at the unblemished coral colonies with halos of colorful fish reaching far beyond the crystalline visibility, while a curious giant green turtle nudged its head up against our video camera. Then, just before our gasping excitement emptied the last air from our tanks, a half-dozen enormous-yet-elegant manta rays tumbled and looped just a few feet from us, feeding in the plankton-rich current.

We surfaced exhilarated and full of questions. What made this place so uniquely rich? Which other extraordinary creatures might thrive here? How must the neighboring islands compare? Could the Phoenix archipelago prove a last remaining wilderness in our besieged ocean? Whatever it took, we knew we had to come back to help unravel these mysteries.

8 PROTECTING PARADISE

EDUARD NIESTEN AND PETER SHELLEY

Can we put a price tag on the Phoenix Islands? The coral reefs, fish species, bird populations, scenic beauty—their value is immeasurable. However, in more practical terms, the government and people of Kiribati depend heavily on money from the sale of fishing licenses. Licensing the right to fish in the waters around the Phoenix Islands funds a variety of educational, health, security, and other fundamental services. Protecting the Phoenix Islands by barring commercial fishing in their vicinity will undeniably safeguard extraordinary natural value; however, giving up the fees from fishing licenses is a tall order for a small, developing island state with few other ways to generate income for its population of 110,000.

At the same time, people all over the world spend time and money to protect the environment—through membership fees to conservation organizations, corporate donations, government budgets, and more—thus exhibiting what economists call a "willingness to pay" for conservation.

Our central task in creating the PIPA, then, was to channel a portion of global willingness to pay for conservation to Kiribati in a way that makes up for the potential loss of revenue from fishing licenses. Under this scheme, instead of letting commercial fishers pay for the right to harvest in the waters around the Phoenix Islands, Kiribati would allow the world to pay for the right to protect this unique ecosystem. In 2002 Minister Tetabo Nakara coined the phrase "reverse fishing license" to describe this proposition—a phrase that perfectly captures an arrangement in which the world makes it financially viable for a developing country to establish one of the largest marine protected areas on Earth.

Before we could make such an arrangement possible, we first needed to answer several complex questions. How much money would be needed to make up for lost fishing license fees? What were the legal mechanisms needed in Kiribati to ensure that the PIPA was duly authorized under Kiribati law and protected by a comprehensive and enforceable regulatory structure? What mechanism

Fishermen on Orona
preparing longlines.
(Cat Holloway)

could guarantee that this money would be available to the government of Kiribati on a timely, predictable basis? How could those who provided the financing for the reverse fishing license—the potential conservation investors—be assured that the funds would achieve their intended purpose? To answer these questions, we drew on a range of economic, financial, and legal expertise and benefited from a unique collaboration among the government of Kiribati, the New England Aquarium, and Conservation International. What follows is a brief account of the contractual and legislative challenges facing the PIPA partners and how those challenges were met.

Valuing the Phoenix Islands

What was the protection of the Phoenix Islands worth? To answer that question, the PIPA partners commissioned a natural resource valuation of the Phoenix Islands. The crucial part of this valuation was assigning a reasonable dollar amount to the fishing license fees generated by the waters around the Phoenix

Islands, but the study also examined the value of other resources, such as sea cucumbers and coral reefs. The resulting analysis required a tremendous effort and resulted in a groundbreaking study—the first at this scale for a marine protected area in a developing Pacific island nation.

Fishing Revenues

The hardest task we faced was calculating the appropriate compensation to the Kiribati government for lost fishing revenues. Kiribati receives fishing license fees through two types of arrangements. First, the fleets of other countries—known as distant water fishing nations, or DWFNs—pay annual access fees for the right to harvest in Kiribati's territorial waters as well as in its extensive exclusive economic zone (EEZ). Nations generally have territorial seas that extend 12 nautical miles (22 km) from their shores. Beyond these territorial seas, all nations recognize an additional exclusive economic zone extending 200 nautical miles (370.4 km) from shore, which grants sovereign jurisdiction over all resource development activities, like fishing, to the adjacent nation. Because of the way these zones are measured, Kiribati has one of the largest EEZs in the world. Second, Kiribati is among a group of Pacific island countries that receive payments under a 1987 fishing treaty with the United States that is renegotiated every ten years; the current treaty runs through 2013. Together, these fees contribute about one-third of Kiribati's annual government revenue.

To determine an appropriate fee for the reverse fishing license, we first needed to estimate the income that fishing licenses brought in from the PIPA. But because of the nature of fish as a resource, this calculation proved challenging. Schools of fish are in constant motion. Tuna, for example, follow fluctuating food sources, currents, and ocean temperature gradients, paying no attention to territorial boundaries. As a result, the catch from any one area within the larger Kiribati EEZ, such as the PIPA, fluctuates from year to year, as does the total tuna catch in Kiribati. Between 1997 and 2006, the Phoenix Islands' share of the total national tuna catch ranged from roughly 4 percent to 43 percent, with an average of just under 17 percent. Thus, in some years, the proportion of the total tuna catch from Kiribati waters that comes from the Phoenix Islands' EEZ is minor, but in other years the Phoenix Islands' EEZ contributes a significant percentage of total landings.

A further difficulty arose because the PIPA comprises only a portion—close to 50 percent—of the Phoenix Islands' EEZ, and existing catch data provided no way to tell how much of the reported catch from the Phoenix Islands' EEZ came from within the smaller PIPA zone. Moreover, it was impossible to determine how, or even whether, the total Kiribati catch might decline if all or part of the PIPA were placed off-limits to fishing; the commercial fleet operating outside the PIPA would likely still catch a portion of what otherwise would have been caught inside the restricted area. Indeed, by protecting tuna-spawning areas within the PIPA, the closure could conceivably serve to improve yields in neighboring waters.

Tuna follow fluctuating food sources, currents, and ocean temperature gradients and, as a result, the catch from any one area within the larger Kiribati EEZ, such as the PIPA, also fluctuates from year to year (*pages 110–11*). (Brian Skerry)

Inshore Fisheries

This commercial fishing vessel is using the purse seine method to fish the waters of the Phoenix Islands. Barring commercial fishing within the PIPA will help safeguard this ocean wilderness; at the same time, Kiribati has few other ways to generate income for its population. (Cat Holloway)

Island and coastal nations can sometimes earn income from their inshore fisheries and reef resources. In contrast to high-seas fishing fleets that stay at sea for weeks and months at a time, inshore fisheries typically use small open boats based along the shoreline that go to sea daily and return to port each nightfall.

The team compiling the Phoenix Islands valuation looked at existing and potential inshore fisheries in the PIPA to see whether they ought to be taken into account. We found that although some commercially valuable species are found in the waters just off the Phoenix Islands, the remoteness of the islands, and the resulting high cost of transporting the catch to market, makes them unprofitable to harvest. As a consequence, inshore fisheries did not figure into the final number on which we calculated the reverse fishing license fee.

Another issue raised in the valuation was the possibility that instead of granting licenses to foreign fishing vessels, Kiribati might one day develop its own fishing fleet. Domestication of the fishing fleet could increase the economic benefits derived from exploitation of the nation's resources, thereby changing the amount on which the reverse fishing license is calculated. However, despite long-standing interest in this prospect, a domestic fishing industry has yet to emerge, and it is unclear when it might do so, so it was not included in the present valuation.

How Much Is a Coral Reef Worth?

The team also considered whether the valuation should include the economic value of the coral reefs surrounding the Phoenix Islands. Placing a value on a coral reef is a daunting assignment; still, economists can use a number of approaches to try and assign a figure to its value. (For more information about the dollar valuation of coral reefs, see the World Resources Institute Report in the references.) They can consider the published values of other coral reefs, damages and fines paid by vessels that have grounded on coral reefs, and costs incurred by cities for shoreline protection, as well as possible royalties from pharmaceutically important species and the value of biodiversity itself. Coral reefs are now recognized as the "rainforests of the sea," as they are nurseries for innumerable marine species, potential sources of new medicines, and important regulators of global climate. These various approaches suggest that the value of the Phoenix Islands coral reefs could amount to billions of dollars. In the end, however, the partners agreed that the value of the reefs did not belong in the formula for the reverse fishing license fee because Kiribati would not be surrendering or compromising any of that value by establishing the PIPA. In fact, Kiribati seems likely to benefit as the PIPA preserves and enhances the value of its coral reefs, adding to the country's natural capital. Kiribati voluntarily closed all commercial coral reef fisheries in the Phoenix Islands with no compensation in 2002.

A fisherman holds up his catch. Most I-Kiribati fish to feed themselves and their families. The small-scale native fishery exports pet fish for the aquarium trade, some artisanally landed shark fins, and seaweed. (Cat Holloway)

What to Include?

Leaving inshore resources and coral reefs out of the final compensation formula doesn't mean that they are without value. They are tremendously important as the basis for global willingness to pay for conserving the Phoenix Islands. However, after a year and a half of research, analysis, and discussion, the PIPA partners agreed that potential lost fishing license revenues alone should serve as the basis for reverse fishing license fees. They settled on a methodology and a number of simplifying assumptions to derive an initial estimated valuation of roughly $1.7 million to $3.25 million a year.

The team also had to consider the costs of managing the PIPA. The PIPA will be operated under a management plan that includes monitoring, research, enforcement, and zoning. Zoning will be particularly important, since this will determine where and how various activities such as tourism, recreational diving, and sustainably managed fishing will be permitted. The PIPA partners determined that a management budget of $300,000 per year would be a bare minimum.

The Funding Guarantee for Kiribati

The second major question faced by the PIPA partners was how to ensure the payment of annual fees covering the reverse fishing license and the new management costs to the government of Kiribati. In time the PIPA will generate revenues through licensing fees for activities such as offshore fishing and tourism to the islands, new domestic employment, and goods and services purchased in connection with tourism and scientific research. It may also generate local revenues through artisanal uses of resources, such as small-scale fishing and shellfish collection. However, for the foreseeable future, revenues from such activities will certainly not be enough to make up for the loss of potential income from commercial fishing licenses. If it was to set aside the Phoenix Islands as a permanent marine protected area, Kiribati had to be assured that the reverse fishing license fees would be forthcoming and predictable, rather than facing the risk of annual revenue shortfalls. The best way to guarantee these annual payments was through an endowed trust fund dedicated to the PIPA.

An endowment is a permanent trust fund from which only the investment income is used. In this way the original investment, or capital, of the trust remains intact, guaranteeing an annual stream of revenue in perpetuity. The PIPA endowment had to be large enough to ensure that the annual investment income could cover the reverse fishing license fees and the costs of managing the protected area as well as supporting the operations of the trust fund itself.

After many months of research and analysis, team members from the New England Aquarium and Conservation International flew to Tarawa for several days of consultation with the government of Kiribati, particularly the Ministries of Environment and Fisheries. During the first day, the meeting felt like a negotiation, with one side trying to marshal arguments to minimize the reverse fishing license fee and the other seeking to maximize government revenue. At this stage, it was hard to see how the PIPA partners on opposite sides of the table could converge on an amount for the annual payment—where in the wide range of $1.7 million to $3.25 million the figure should fall. It seemed that the PIPA endowment could not move forward without that number, which meant that the whole project might be stuck.

To start the second day of discussions, the partners took a step back to remind each other that, instead of a negotiation, this exercise would be better seen as a collaborative effort to devise a workable plan for financing the creation of what was at the time the world's largest marine protected area. We also noted that the financial analysis involved several unknowns, so that the numbers involved subjective judgment rather than precise calculation, and that we were free to revisit our assumptions over time when periodic reviews of the program took place. This meant that precise agreement on the numbers was unlikely and unnecessary. We all wanted the best deal possible for the people of Kiribati, but we also needed to be realistic about how much funding we would be able to raise together.

The typical approach to figuring out how much money needs to go into an endowed trust is to estimate the annual financial need and then multiply that amount by twenty. In effect, this approach assumes that the annual yield on the endowment will be 5 percent; in other words, that every $100 in the endowment will yield $5 in investment income to spend each year, forever.

With this new perspective, the partners decided to approach the overall question of a suitable compensation arrangement from a different angle. Based on our collective experience, we figured that an ambitious, yet realistic, initial target for the endowment would be $25 million. Working backward from this endowment amount (assuming an annual yield of 5 percent and taking into consideration management costs and the cost of operating the trust fund), we determined that this amount would be commensurate with the creation of a no-take zone (NTZ) equal to 25 percent of the total area of the PIPA. This would bring the no-take and restricted use zones within the MPA to a total of 28 percent.

This decision might seem like a drastic compromise on the part of conservation, as only 28 percent of the PIPA would be completely off-limits to commercial fishing. However, this 28 percent would completely protect seven of the eight Phoenix Islands and associated coral reef systems (Kanton would still have a minor artisanal fishery), two submerged reef systems, and several key seamounts. Furthermore, commercial fishing in areas outside this no-take zone would be subject to strengthened regulations to ensure sustainable fishing practices. This approach would preserve the attractiveness of Kiribati's overall EEZ to foreign fishing fleets, as it would allow fleets to continue their current practice of fishing in multiple harvest areas. In addition, a 28 percent NTZ would be roughly consistent with the precedent set by Australia's Great Barrier Reef Marine Park, which has a 33 percent NTZ (in fact, the initial Barrier Reef NTZ was only 4.6 percent). It would also be consistent with IUCN's target of protecting 20 to 30 percent of total ocean area as NTZs by 2012.

While raising $25 million for the PIPA endowment would be no trivial task, the partners recognized that if the endowment could be grown beyond that target, the NTZ could expand proportionally. In addition, they decided to track trends in harvest and license revenues from the commercial fishing zone of the

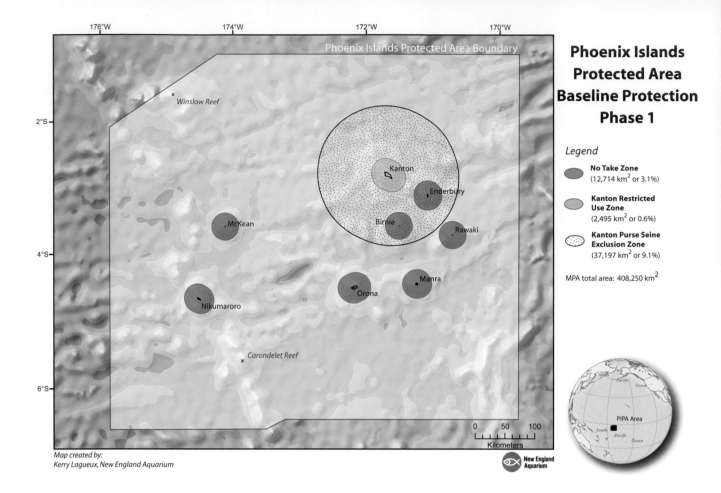

Phoenix Islands Protected Area Baseline Protection Phase 1

Legend

No Take Zone
(12,714 km² or 3.1%)

Kanton Restricted Use Zone
(2,495 km² or 0.6%)

Kanton Purse Seine Exclusion Zone
(37,197 km² or 9.1%)

MPA total area: 408,250 km²

Map created by:
Kerry Lagueux, New England Aquarium

New England Aquarium

Phoenix Islands Protected Area boundaries as of 2008. The NTZ comprises a range of 12 nautical miles around each of seven of the eight islands. There is also a 12-mile restricted-use zone around Kanton and a 60-mile radius around that island within which purse seine fishing is not allowed. (New England Aquarium)

PIPA EEZ with greater precision so that the relationship between the endowment level and the size of the NTZ could be refined with more precision over time.

Creating the Legal Framework

The success of the Phoenix Islands Protected Area concept also depended on the adequacy of the legal framework for managing the trust and enforcing conservation goals. No third party, public or private, would invest in the PIPA without this legal scaffolding in place. The structure we settled on for the PIPA was heavily influenced by the Kiribati government's desire to maintain sovereignty and control over its natural resources while at the same time ensuring protection of the long-term conservation interests of the New England Aquarium, Conservation International, and other third-party investors in the conservation trust fund.

The key to making it work was creating a trust that operated under Kiribati law, structured to be fully transparent and independent of Kiribati government control. The trustees would negotiate a conservation contract with the government for the management of the PIPA that would meet all these strictures and satisfy the legitimate interests of the various stakeholders. In order to create such a trust, we needed new legislation.

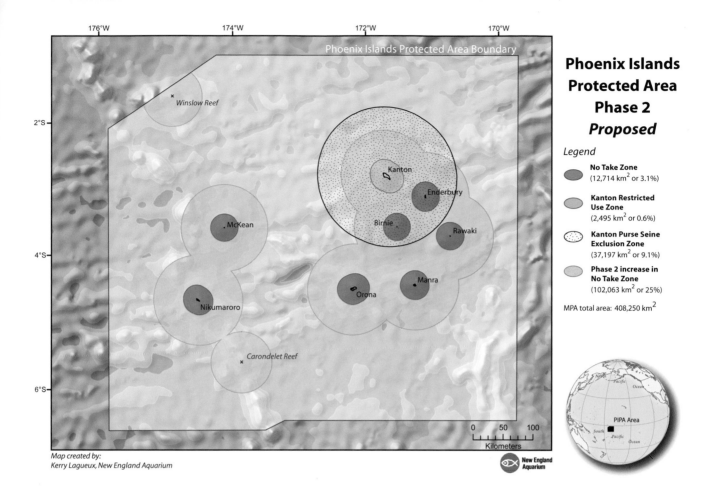

Phoenix Islands Protected Area Boundary

**Phoenix Islands
Protected Area
Phase 2
*Proposed***

Legend

No Take Zone
(12,714 km^2 or 3.1%)

**Kanton Restricted
Use Zone**
(2,495 km^2 or 0.6%)

**Kanton Purse Seine
Exclusion Zone**
(37,197 km^2 or 9.1%)

**Phase 2 increase in
No Take Zone**
(102,063 km^2 or 25%)

MPA total area: 408,250 km^2

2°S

Winslow Reef

Kanton

Enderbury

4°S

McKean

Birnie

Rawaki

Manra

Orona

Nikumaroro

6°S

Carondelet Reef

0   50   100
Kilometers

PIPA Area

Map created by:
Kerry Lagueux, New England Aquarium

New England
Aquarium

One possible spatial configuration of Phoenix Islands Protected Area NTZs, which would expand the area protected as of 2008 by 25 percent. This additional protection will come into force when the capitalization target of $25 million is reached. (New England Aquarium)

By 2005 the government and conservation partners recognized that a new legal-enabling framework would also be needed to create the PIPA itself. The existing laws of Kiribati were not adequate for either the legal recognition of a special status for the PIPA or the management, administration, or enforcement of the measures that would ensure that it accomplished its conservation and biodiversity protection goals. The conservation laws then in existence, largely remnants from the British colonial era, protected some of the country's terrestrial and lagoon resources, but they were out-of-date and inadequate for the PIPA's purposes. Even though aspects of Kiribati's environmental laws had been substantially updated as recently as 1999, they had not been enacted with a major initiative like the PIPA in mind.

Accordingly, with the leadership of the Ministry of Environment, Lands and Agricultural Development (MELAD) and the Office of the Attorney General, a major revision of Kiribati laws was undertaken, beginning in 2005 and culminating in the passage of the Environmental (Amendment) Act 2007, which took effect on September 4, 2007.

Under this new authority, then MELAD Minister Nakara, acting on the advice of the Kiribati cabinet, formally enacted the new legal framework for the PIPA, known as the Phoenix Islands Protected Area Regulations 2008 (2/7/2008). The new regulations defined the land and marine boundaries of the PIPA, approved the nomination of the PIPA for inscription on the list of World Heritage sites, and identified the PIPA as a Category 1b Wilderness Area under the definitions provided by IUCN's 1994 *Guidelines for Protected Area Management Categories*. They also created a new interagency management committee and established the scope of the PIPA Management Plan. They further established that all persons and all public authorities undertaking any activities in the PIPA are subject to the Environmental Act of 1999, the PIPA regulations, and the PIPA Management Plan.

The PIPA partners next turned their attention to the matter of establishing the conservation trust. They were comfortable with the idea of establishing a conservation trust in Kiribati under domestic law, as the country had had an exemplary history of managing its Revenue Equalization Reserve Fund (established in 1956 and funded by revenues from phosphate mining on some of the islands) and the Kiribati Provident Fund (established in 1977). There was, however, no explicit statutory mechanism for creating a tax-exempt conservation trust or conservation trust fund that had all the features critical to the partners and conformed to international best practices for conservation trust funds.

The partners resolved to remedy that deficiency quickly. Leaning heavily on tax and corporate legal expertise provided by the Boston-based law firm Ropes & Gray LLP, they drafted legislation establishing the Phoenix Islands Protected Area Conservation Trust and circulated it among the parties. There were two overriding considerations that shadowed the drafting process. First, the trust had to be a Kiribati corporation with strong government representation on its governing board, but without government control of its operations and decisions. Second, the trust had to be set up so that the United States would recognize its charitable status under U.S. tax law.

Both objectives were accomplished. After two years of development and negotiation, the PIPA Conservation Trust was established in May 2009. The final legislation established the trust under Kiribati law and domiciled the corporation in Kiribati. While the primary purpose of the trust is to provide financial support for the PIPA, the trust's corporate purposes were broadened in order to allow the trust to be used for other conservation purposes in Kiribati at the trust's option. The government realized that the trust could help secure and administer outside conservation funding that could be used to advance adaptation to climate change, conserve natural areas beyond the PIPA boundaries, or pursue other conservation or environmental protection initiatives. The trust is dedicated to charitable purposes and is therefore exempt from income taxes, customs duties, and similar taxes.

The PIPA Conservation Trust is governed by a core board of three directors, appointed by the government of Kiribati, the New England Aquarium, and Conservation International. The founding directors are authorized to appoint additional directors, expanding the board as they deem appropriate. The legislation provided specific protections to ensure that the government of Kiribati would always be in a minority position on the trust board, including a requirement that a majority of the directors come from outside the government. In addition to a number of conventional measures establishing the corporate structure, powers, and duties of the trust, the legislation provided for the appointment of an executive director by the board. Special majority voting rules were also put in place to ensure that all actions related to the trust assets reflect the conservation and charitable objectives of the trust.

Fundamental to the project's success was the clear understanding that the trust would have no oversight role in developing or implementing the PIPA Management Plan. While the trustees are entitled to comment on the plan and any amendments to it, their role is limited to negotiating a conservation contract with the Kiribati government, accepting or rejecting the government's management plan as part of those negotiations, and ensuring that the reverse fishing license fee and the costs of managing the PIPA are paid to the government. If the trust is not satisfied with the terms of the management plan or its implementation, then under the terms of the contract, it will not be obligated to make payments to the government or renew the contract when it expires. This contractual, readily enforceable arrangement and the lack of government control of the voting structure of the PIPA Conservation Trust ensure that the government's sovereignty over activities in the PIPA is not compromised while assuring outside conservation investors in the PIPA that their funding is being applied precisely for its intended purposes.

The first meeting of the PIPA Conservation Trust was held in March 2010 in Tarawa. The founding board members have been appointed, and bylaws have been drawn up and approved by the new board. Attorneys for the partners have also begun drafting the conservation contract that will be executed between the trust and the government of Kiribati. As noted above, performance under the contract will be measured by the government's compliance with the provisions of the PIPA Management Plan and, when the assets in the trust fund are sufficient to produce an appropriate income stream, the government's actions to prohibit deepwater or other fishing in the designated PIPA no-take zones.

While conservation contracts have been used successfully for some time to protect rainforests and other terrestrial resources, the PIPA Conservation Trust represents the first time this mechanism has been applied to ocean resources. Compensating developing nations for the conservation of their natural marine resources is a reasonable way to secure the benefits of those resources while providing support for the manifest social needs of the host country. Around the world, the eyes of governments and conservationists alike are on the PIPA experiment. Much is at stake, both for the Phoenix Islands and the people of Kiribati and for the success of conservation efforts around the globe.

Expedition Diary: A Legal Education, Kiribati Style, Tarawa, 2004

PETER SHELLEY

In December 2003, Rockland, Maine, was in the grip of a bitter winter storm. Wind-whipped sleet and rain pelted my window at an almost horizontal angle. I was sitting at my computer gazing out at the weather when the phone rang.

It was Greg Stone, and his call offered respite from the overwhelmingly gray day.

"How would you like to do some legal work for a Pacific Ocean island nation?"

So incongruent was his question with my storm-besieged state of mind that I was dumbstruck.

"I am doing this terrific project working closely with a Pacific islands country to save what might be one of the last intact archipelago reef systems in the world," he explained. "I need someone who can do the legal and policy work that will be necessary."

"I want you to go with me to Tarawa, which is near the equator," he continued, "and spend some time with people there to figure out how we can make this happen," Greg explained. "You in?"

A gust violently rattled my office windows as I looked out over Rockland Harbor, trying to imagine sunshine and palm trees. I couldn't. "Sure," I answered, not having the slightest idea what I was getting myself into, but seriously motivated by the prospect of sunshine and warmth.

"Great," Greg shot back. "I'll give you a call when we work out all the details." Then he hung up.

The "details" worked out to my traveling to Tarawa for a month in June 2004. The goals for my time there were to develop direct relationships with the government officials with whom we would be working, to gain a better understanding of the legal and regulatory culture in Kiribati, and to do some preliminary research on Kiribati law so that we could understand better what legal tools were already in place and which we would have to develop.

I emerged from the plane at Tarawa's Bonriki International Airport after an interminable flight, blinking into the blistering sun of an equatorial summer day. Overhead, the brilliant blue sky was decorated with towering cumulus clouds forming for the afternoon rain shower. I still had no idea what I was getting myself into.

My legal world in New England is driven by e-mails, computer-assisted legal research, and immediate fingertip access to virtually any document. When I arrived in Tarawa, however, there was no organized system of lawbooks or regulations either in the agency I was working with or in the National Library. The only complete set of laws seemed to be in the Office of the Attorney General, where it was in heavy use. In addition, Kiribati law was a patchwork of ordinances dating from British colonial days and new laws dating from Kiribati independence in 1979. Kiribati is a recently formed democratic nation that operates under the parliamentary system. Kiribati operated very well under the rule of law—indeed, it managed sophisticated international relations under that law—but it wasn't easy for a foreigner to maneuver the reefs of Kiribati legal culture. Just getting a copy of a single ordinance or regulation could turn into a full-day exercise.

I spent the next four weeks meeting with numerous government officials, learning about Kiribati law and the parliamentary process of enacting new laws, and evaluating the degree to which existing Kiribati law could support the sort of conservation program that would satisfy third-party donors and other key program participants, such as UNESCO's World Heritage program. This chapter outlines the results of that investigation and the follow-up actions Kiribati and the PIPA partners took.

Today, thanks to some wonderful pro bono work by a young Australian lawyer, Marcus Hipkins, all Kiribati laws and ordinances and many of the environmental regulations have been scanned and are now available in searchable electronic format. That makes finding a particular law much easier. I am glad, however, that I had to learn the legal ropes the hard way through a more people-intensive approach. I came away not only with a working understanding of Kiribati law, but also with a deep respect for the people for whose benefit that law was operating.

9 THE FUTURE OF THE PHOENIX ISLANDS

GREGORY S. STONE, DAVID OBURA, SUE MILLER-TAEI, AND TUKABU TEROROKO

Expedition Diary: Meeting in Tarawa, 2001

GREGORY S. STONE

The Air Nauru jet roared across the flat Tarawa atoll, landing on a bumpy runway where children play between flights. Back in 2001 the terminal was the size of a house, open to the air, with a thatched roof. When the large plane landed twice each week, it was an event, drawing locals who gathered along the chain-link fence to check out the new arrivals. As soon as the door of the aircraft opened, the blast of hot air hit me, and I was once again reminded I was near Earth's equator.

Tarawa's single road winds around the narrow atoll, a strip of land no wider than a few hundred yards studded with coconut palms, pandanus trees, breadfruit trees, and other tropical plants, with the lagoon on one side and the beautiful blue Pacific Ocean on the other. The highest point is only 4 or 5 feet (1.2–1.5 m) above sea level. With sea levels rising, it is estimated that Kiribati and many of the other forty-two small island states will also be submerged or seriously impacted within a century.

A government car had been sent to meet me. I held tight to the handgrips as my driver swerved to avoid the many deep ruts in the road. It was blast-furnace hot, and the vehicle air conditioner, though trying, could not keep up; I was tempted to open a window instead. Many of the homes we passed were simple huts made of logs lashed together with twine and thatched roofing. The floor of each small dwelling was raised off the ground to allow air to circulate and keep rats at bay. There was also an occasional gas station or small store selling drinks, snacks, and cigarettes, and a church every few miles. Skinny dogs, mangy cats, and chickens roamed freely, while pigs were tied in pens outside many homes. Tropical birds of a species I had never seen before circled above the water and treetops, making their calls. As we passed, I could see people escaping the midday heat, napping with colorful but faded sarongs wrapped around them. An elderly woman carried water to an open cooking fire as smiling children, naked and nearly naked, turned to track the movement of my car with their eyes as we drove by.

The closeness of Kiribati's national culture to the ocean is expressed in its songs and dances, which glorify and emulate the motions of seabirds, fish, and other sea creatures and tell stories of men and women living with the sea. (Tukabu Teroroko)

Forty minutes later, we reached the capital of Kiribati, South Tarawa, best known to outsiders as the location of the Battle of Tarawa during World War II. Here, the houses were the modernized, square boxes of concrete block that are typical throughout the developing world: dusty, partly painted. Cooking smells of fish and roast pig wafted from open doorways. Puffs of blue oil smoke erupted out of the vehicle exhaust pipes of small minibuses and mopeds. These are the sights and sounds of life as it is lived in many parts of the world where people subsist on the lowest of incomes.

At the Ministry of Fisheries, I was met by David Obura and Sangeeta Mangubhai. In a room with a struggling air conditioner, the three of us presented a slide show and science report to the ministers of Fisheries and Environment—well-educated, middle-aged men in ties—showing them magical underwater scenes of hundreds of sharks, lavish, colorful corals, dense clouds of reef fish, and countless birds nesting on the islands. They were as amazed as we had been at the untouched, museum-like quality of the Phoenix Islands reefs, in sharp contrast to the reefs closer to their population centers, which were all overfished and degraded by coastal development. I could tell that seeing these thriving reefs meant a great deal to the Kiribati people in the room because of their society's ancient connection to the sea, and because the isolation of the Phoenix Islands means that almost no I-Kiribati gets to visit them.

Kaburoro Ruaia, one of the officials in the room, informed us that we were the first foreign scientists who had ever bothered to come to Tarawa to explain in person what we had done and what we learned while conducting research in Kiribati waters.

A Cultural Partnership

Our meeting on Tarawa in 2001 marked the beginning of a remarkable cultural conversation and partnership between Western science and conservation institutions and the people of Kiribati. The natural treasure trove that is the Phoenix Islands has a powerful impact on everyone who is privileged to see it. It is something like a conversion experience. Images of the healthy schools of fish, colorful reefs, and islands teeming with rare nesting birds have the power to reshape careers, inspire new generations, and change minds. Seeing these resources in their natural state led Kiribati government officials to understand the value of preserving and maintaining them for the future.

Although the sea is part of the past and present daily life of the I-Kiribati people, most have never had the chance to really observe and appreciate the variety and splendor of the underwater marine life. Few I-Kiribati are trained or equipped to snorkel, let alone to scuba dive. The presentations we made at the meeting truly inspired the people who saw them.

The Phoenix Islands have turned marine biologists into passionate conservationists and government officials of tiny Kiribati into pioneers of conservation law. Now, a decade on, when the stakes for the planet could not be higher, the Phoenix Islands have a chance to inspire more people to make the personal commitment to save our imperiled ocean.

When the Phoenix Islands were rediscovered in 2000, they were one of the last oceanic coral archipelagoes in the world to be unexplored by modern science. Greg Stone was not yet a fully fledged conservationist, devoting most of his professional time and energy to exploration and original scientific research. The Phoenix Islands changed all that. Conservation is now Stone's full-time job in his position as Executive Vice President and Chief Scientist for Oceans at Conservation International. The Phoenix Islands showed Stone what the ocean may have been like a thousand years ago, before humans began to leave their mark

Once fully endowed, the Phoenix Islands Protected Area Conservation Trust will provide Kiribati with income to fund services such as education. (Brian Skerry)

(Randi Rotjan)

One of the biggest challenges facing wildlife manager Aranteiti Tekiau during his summer fellowship at the New England Aquarium was actually getting to Central Wharf each morning from his temporary lodgings in Boston's Jamaica Plain neighborhood. Every morning he rehearsed the commute in his head before setting out. "In Tarawa," he says, "there is a single road, and it's impossible to get lost."

Tekiau represents a new generation of I-Kiribati government employees working to document and preserve the unique biological heritage of their island nation. After receiving training in fisheries management at the University of the South Pacific and managing a fishery in Marakei for two years, Tekiau journeyed halfway around the world to master the most up-to-date coral reef monitoring techniques and bring them back to Kiribati. Coral reef monitoring allows scientists to assess the diversity and health of reefs and to develop effective strategies for managing them in the face of human-generated stresses from pollution and overfishing to climate change.

While he was in Boston, Tekiau learned techniques and procedures for monitoring reef organisms in the labs of Randi Rotjan at the New England Aquarium and Les Kaufman at Boston University. Tekiau stood outside the aquarium's 200,000-gallon, four-story Giant Ocean Tank as a diver laid a transect 40 feet across the tank, then spent hours learning to rapidly and accurately count, identify, and gauge the sizes of fish as they swam by, twenty-five or thirty at a time.

Tekiau also learned to separate symbiotic algae from their coral hosts, using such unexpected tools as a dental Waterpik and a blender. When the algae were finally "spun down" and isolated, Tekiau got his first look at them under the microscope and learned how to use their green-brown color to determine their health.

Tekiau also visited the Large Pelagics Research Center to learn about ongoing monitoring of tuna and sea turtles from researcher Molly Lutcavage, and he visited the Woods Hole Oceanographic Institution to discuss blue-water species. All of this amounted to what Rotjan calls "rigorous training in the global standard of coral monitoring." Now Tekiau is back in Tarawa, putting his new skills to work and sharing his knowledge with fellow wildlife managers working to protect and preserve the reefs of Kiribati.

on nearly every place on Earth. Even more importantly, they made him realize what the ocean could be like in the future if we conserve and protect it now.

David Obura found the Phoenix Islands just as powerfully inspiring. They renewed his vision on his path of combining science with conservation. For the first ten years of his career as a coral biologist, he had had a myopic view of what constituted a healthy coral reef. The Phoenix Islands provided his first view of pristine reefs, something he had not seen in his studies or work.

In Africa tens of millions of people fish to eat and earn their livelihoods from the environment. They have a right to more than just survival; they have a right to prosper. Thus efforts to conserve Africa's last natural places and to maintain the health of its altered ecosystems are hampered by competing pressures and influences, which force a desperate triage among dismal options.

The Phoenix Islands and the culture and pride of the I-Kiribati offered David a chance to dream of what might be possible to achieve in reef conservation—and to work with the Kiribati government and with international colleagues to take the steps to fulfill that dream.

For government officials in Kiribati and conservationists throughout the Pacific, the Phoenix Islands offered a rare confluence of conditions that made a bold stroke of conservation possible. Isolation had thus far spared the Phoenix Islands from the toll of human impacts. Except for the tiny settlement on Kanton, there were no resident communities whose livelihoods and self-determination might be threatened by conservation. Among the I-Kiribati, there was a single national culture that valued and respected the ocean and its small islands, a culture not (yet) overly influenced by rampant global consumerism. Members of the Kiribati government—from the officials working at ground level through the president—found something of value in the dream of the protected area and worked together over the years to make it happen.

Tukabu Teroroko is one of those in Kiribati who is committed to the PIPA. As the inaugural PIPA director, he has led the day-to-day development of the PIPA since its establishment. He has won many over to the PIPA cause, most recently at the UNESCO World Heritage meeting, where he worked incredibly hard as part of the Kiribati delegation, with the support of fellow Pacific and small island states, to secure PIPA's World Heritage site listing. Today the PIPA is the world's largest and deepest World Heritage site. Teroroko is affectionately known as "Mr. PIPA."

Sue Miller-Taei, Pacific Islands Marine Director for Conservation International, shares the workload of the PIPA. Her contributions include facilitating the management plan process and the World Heritage site listing of the PIPA. Being part Samoan-Kiwi with family links throughout the South Pacific, she lives in both Western and Pacific island worlds. This experience helped her develop not only the design of the PIPA, but more recently the Pacific Oceanscape initiative. With four young children, she is the only one of the authors of this chapter who has yet to visit the PIPA, since the trip requires at least a month. She looks forward to visiting the PIPA in the future, but in the meantime believes that millions of people, particularly those living in urban environments, need to know about, believe in, and support wilderness areas like the PIPA, not just for the planet's well-being, but for their own spiritual well-being, even if, like Sue, they can't visit these areas . . . yet.

The Phoenix Islands as Inspiration

Today most of the world's ocean is in serious decline, especially coral reefs, which are overfished and dying due to coastal development, erosion and runoff, pollution, and global warming. Back in 2000 the Phoenix Islands' coral reefs were the most pristine ever found. Virtually uninhabited, the islands had been protected by their remoteness, seeming to exist outside of time, and their reefs had been spared the overfishing and destruction that had ravaged neighboring reefs in the Pacific and elsewhere.

The ocean is the planet's largest habitat, covering more than three-quarters of its surface, yet most of this environment is unexplored. We know more about the surface of Mars than we do about the seafloor of our own planet.

We must take care of our ocean. The fact that only about 1 percent of the ocean is protected as this volume goes to print (compared with 12 percent of the landmasses) should be a clarion call for ocean conservation. We need to ramp up our protection of the ocean and create more large marine protected areas like the PIPA, the Great Barrier Reef Marine Park, and Papahānaumokuākea Marine National Monument, to name a few. There has never been a more important time than the present to do so.

Pacific nations have taken a leadership position in global marine conservation. In 2009 President Anote Tong of Kiribati presented a concept to fellow member nations of the Pacific Islands Forum, proposing a "Pacific Oceanscape" initiative to protect the world's largest ocean. In 2010 the fifteen-member

SIDEBAR 9.2: Surveying Seamounts · *Gregory S. Stone*

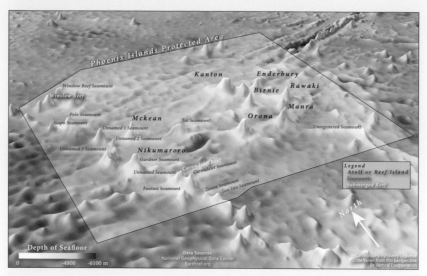

(New England Aquarium)

A priority for exploration and conservation within the PIPA is its more than fourteen underwater mountains, or seamounts. Rising up from the ocean floor, thousands of meters tall, these submerged volcanic peaks are found throughout the world's oceans, but until the last five years scientists didn't realize that seamounts were one of the most common biomes in the marine environment, or that they were biodiversity hotspots rivaling rainforests and coral reefs in their species richness. They may be especially important as breeding grounds for species that spend the rest of their lives elsewhere in the ocean and as refuges for species under pressure.

Around 2004, when I cowrote a chapter about seamounts for *Defying Ocean's End: An Agenda for Action*, scientists had identified 300 seamounts around the world and had thoroughly explored about 200. As of 2010, with the conclusion of the decade-long Census of Marine Life (COML), the estimate now stands at a staggering 100,000 seamounts, covering roughly 11.1 million square miles (28.8 million km²), or an area larger than the continent of South America. These previously unmapped seamounts were revealed with the aid of satellite altimetry, which allows accurate measurements of variation in the seafloor and statistical analysis.

In 2008 I had a chance to visit a seamount during a New England Aquarium expedition to the Sea of Cortez. I ventured below 1,000 feet in the submersible *DeepSee* to observe the marine life around El Bajo, a seamount off the coast of Mexico renowned in the past for its vast congregations of hammerhead sharks and as the location of Peter Benchley's 1982 novel *The Girl of the Sea of Cortez*. As we steered the *DeepSee* past the summit of El Bajo, schools of redfish and amberjacks swam past the Plexiglas bubble that was our window on this hidden world. As we passed, garden eels peeked out from their sandy burrows, but we saw no hammerheads. Then we came upon a "ghost net"—an old seine that had wrapped itself around one of the seamount's peaks. Here was an unmistakable sign of the overfishing that had reduced the once-plentiful sea life of El Bajo to a fraction of its former abundance.

Are the seamounts of PIPA safe from overfishing? Officials from IUCN believe that fully 40 percent of the yellowfin tuna caught in the Pacific are caught by long-lining fishing vessels on or around seamounts. There is a powerful economic incentive for fishing fleets to target these species hotspots.

The establishment of the PIPA and its inscription on the list of World Heritage sites should spare its seamounts from such a fate, but monitoring and policing of the PIPA waters will be necessary to keep its seamount ecosystems intact until they can be sampled, surveyed, and mapped. Amazing discoveries about seamounts and the life they support remain to be made. In 2008 COML researchers discovered brittle stars feeding by the tens of *millions* on the summits of a chain of seamounts between New Zealand and Antarctica. What surprises might await us on the Phoenix Islands' submerged peaks? For now, we can only wait and wonder.

nations' leaders, including President Tong, met in Vanuatu and endorsed the Pacific Oceanscape Framework. Covering nearly 25 million square miles (40 million km²) of ocean and island ecosystems—the size of Canada, the United States, and Mexico combined—the Pacific Oceanscape unites Pacific island states in a survival campaign for ocean conservation and management in the twenty-first century.

To design the Pacific Oceanscape, stakeholders were recruited from fisheries, universities, conservation organizations, regional agencies, and governments

The electronic tag being implanted into this bigeye tuna will provide fisheries managers and conservationists with valuable life-history data that will help them manage the fish stocks of the Pacific. (Bruno Leroy)

to create a framework for the long-term, sustainable management of the ocean region that the Pacific island nations depend on for their survival and treasure as their natural and cultural heritage. This innovative and ambitious initiative is remarkable not only for its scale, but also for the united Pacific voice it brings to issues of sustainable development and ocean conservation.

The conversation that began on Tarawa in 2001 marked the beginning of a remarkable partnership. The story of the Phoenix Islands shows how a single small action by a few individuals can grow to enlist the expertise, energy, and passion of people around the world, from ordinary citizens to professionals to policymakers at the highest levels of government. Turning conversations into action requires each of us to commit ourselves anew, dedicating our time, energy, financial resources, and political will to the problems of climate change, overpopulation, and the other forces that affect our beleaguered water planet. It demands that we engage in free and honest dialogue across disciplines, cultures, and oceans. Our children, and their children, can inherit this blue planet only if we make the urgent mission to save the ocean our own and raise our voices along with our hopes.

A view from *Nai'a*'s porthole evokes Earth as seen from space, a reminder that this water planet is our Spaceship Earth. (Randi Rotjan)

10 LIVING BLUE MAKING CHOICES WITH THE OCEAN IN MIND

HEATHER TAUSIG

How do our choices here at home affect ecosystems and societies half a world away? The challenges facing the Phoenix Islands are intimately connected to our own choices and actions every day. Saving places like the Phoenix Islands, on the other side of the world, depends on the way we choose to live our lives close to home.

Even if we live far from the sight and sound of the ocean, we are never far from its reach. As the sea rises and falls with the pull of the moon, so our own lives are ruled, whether we are aware of it or not, by the power of the sea. All life on this planet began in the ocean, and the ocean still makes life on this planet possible, in myriad ways, for us as a species and for every individual.

The ocean shapes every aspect of Earth as we know it. It covers nearly three-quarters of Earth's surface, and as it heats and cools, it creates the large-scale weather patterns we experience on land. Plankton in the ocean produce fully half the oxygen in the atmosphere we breathe. Fish, shellfish, and crustaceans form the backbone of ocean fisheries that support millions of people in coastal communities around the world.

Wherever people live, they have drawn artistic inspiration and spiritual renewal from the sea. Marine plants and animals are being examined in labs the world over as bioprospectors search for new pharmaceuticals to revolutionize the treatment of cancer, diabetes, and heart disease. Reefs and coastlines protect coastal towns from storms and erosion, and estuaries provide crucial water treatment as well as nurseries for economically important fish. These are just a few of the many gifts the ocean gives us every day.

We have not repaid the ocean's generosity. Action and commitment now—at the international, national, and local levels—is what it will take to save the underwater Eden of the Phoenix Islands and the rest of the ocean on which the fate of Island Earth depends.

Ocean Education

Robin Culler's students love sharks. Culler teaches at Green Chimneys, a residential facility for emotionally and behaviorally challenged kids in Brewster, New York. Her class, twenty-six boys aged twelve to seventeen, learned about the crisis caused by shark finning and were moved to do something about it. In 2006 Culler's students formed a club, the Shark Finatics, to "adopt" a shark, and they went on to raise thousands of dollars for a shark conservation charity. In 2009 they wrote to President Barack Obama and their members of Congress urging support for laws to protect sharks. They even wrote their own book, *Our Shark Story*. After four years the club had adopted twenty-one sharks and spread their message of conservation to fans around the world. In 2010 the Shark Finatics were named Ocean Heroes for their efforts on behalf of endangered sharks.

But the story of the Finatics doesn't stop there. Culler began teaching local workshops to continue spreading the conservation message and to raise money for Iemanya Oceanica, a shark charity based in California and Mexico. After they graduate from Green Chimneys, the Shark Finatics take their ocean conservation ethic with them; one former student has begun to help endangered pelicans in the wake of the 2010 oil spill in the Gulf of Mexico. Thus a project that started with the adoption of a single shark has grown to touch thousands of people in countries all over the world, inspiring them to take action to save sharks and other animals.

Ocean education creates ripples. As water is connected from stream to river to ocean, our communities are connected in widening circles, from neighborhood to city to state, from country to continent to hemisphere. The efforts of a small group of impassioned individuals have the power to educate and inspire others to take action, far from that first small ripple.

Share your own passion for ocean conservation with others by offering to lead workshops at your local community center, library, or wildlife sanctuary. Reach out to local groups working to preserve your watershed and help educate others about keeping your local rivers and bays healthy. Find out how you can serve as a docent or guide in a local nature park, or volunteer at a natural history museum or aquarium near where you live.

Home Aquariums with a Conscience

A home fish tank is the perfect way to remember to live blue, as it is a constant reminder of aquatic biodiversity and the importance of healthy aquatic ecosystems. This is one of the reasons why keeping a freshwater or marine aquarium at home is one of the most popular hobbies around the world. But where do all those colorful bettas, tetras, clownfish, and tags come from? They can come from wild fisheries or aquaculture, and each has its place in the responsible sourcing of fish for home aquariums.

Many wild-captured fish originate in developing countries where too many collectors still use destructive techniques like cyanide to collect them. These techniques can lead to high mortality before the fish reach the home hobbyist. Fortunately, less destructive ways of collecting fish are being adopted. In the

Rio Negro Basin of the Amazon in Brazil, members of Project Piaba are working to create a sustainable fishery around the cardinal tetra. Local people harvest the bright blue-and-red fish, earning enough to provide an incentive not to cut down the surrounding forest. In the Philippines the International Marinelife Alliance is working with Filipino officials to end the practice of fishing with cyanide, encouraging collectors to use fine nets to collect fish.

But such efforts can succeed in the long run only if change also happens at the other end of the supply chain, with the consumer. If you have an aquarium at home, there are steps you can take to minimize your own impact. Find and support a home aquarium store that is doing its part to make fishkeeping a sustainable hobby. Ask your retailer if the fish have been caught or grown according to best practices. A responsible seller will stock only fish suited for home tanks and refuse to stock species that are threatened in the wild. It is also important never to release an ornamental fish back into the wild, since most are not native to the regions where they are sold.

The ornamental fish hobby can be a force for living blue if it is practiced responsibly.

Stop the Spread of Invasive Species

You're a responsible boat owner. You know that water hyacinth, zebra mussels, and other invasive species can be spread by careless recreational boating. So, before and after every outing, you wash your boat and all your gear. You carefully remove all traces of mud and plant matter from your motor, boat trailer, and other equipment and dispose of it far from the water. Before leaving the boat launch, you drain all live wells, bilge water, and transom wells.

But did you remember the dog?

In order to prevent the spread of aquatic hitchhikers, you need to clean everything that's entered the water—including the family pet. The same goes for your scuba equipment, floats, duck decoys—anything that has come into contact with the water.

Checking the spread of exotic species doesn't stop at the boat slip. There's a lot you can do to stop the spread of invasives. If you garden, replace aggressive non-native species with native plants. Support native fauna by supplying nest boxes for birds and butterflies and bat boxes for threatened pollinators. Never release aquatic plants or non-native fish like ornamental carp into the wild. Don't transport firewood, which may harbor damaging beetles, and learn how to monitor your local country woods or urban forest for signs of damaging insects.

Don't Just Live Blue, Eat Blue

Fish is delicious, nutritious, and heart-healthy, and for decades we've heeded the advice of medical experts to consume more. During those same years, Westerners have acquired an appetite for sushi and sashimi. But the increase in the popularity of fish in our diet has led to overfishing and the collapse of one

fishery after another as we have pulled tuna, swordfish, and other top predators by the millions from the sea.

Smaller fish have not been safe from our voracious appetites, either. A number of sardine and anchovy fisheries have collapsed as tons of those fish have
ended up not on our dinner plates, but ground into feed for pigs, chickens,
and cows.

To help let overfished stocks recover, we all need to educate ourselves about
sustainable fish choices, whether those fish are wild or farmed, large or small.
At the fish counter or at a restaurant, ask questions about sustainability, and
don't hesitate to try a new, more sustainable variety of fish just because it's unfamiliar. Many supermarket fish counters now carry farmed Arctic char and ti-
lapia as well as wild-caught sablefish, pink shrimp, and mackerel. The manager
should be happy to suggest ways to clean and cook unfamiliar varieties. Don't
neglect the invertebrates: consider clams, mussels, scallops, and squid. (Shrimp
can be a good choice too, depending on how and where they were caught or
farmed.) You can often be twice as kind to the ocean by choosing fish that were
locally caught. Making your selection into a soup or chowder can make a pound
of fish or shellfish protein feed a crowd.

And remember that it's possible to meet all your protein needs—and satisfy
your taste buds—by being a true ecological "consumer" and eating a plant-
based diet a few days a week.

Leave a Smaller Carbon Footprint

You know your shoe size, but do you know the size of your carbon footprint?

Start by using a carbon footprint calculator online to see how much carbon
dioxide you are adding to the atmosphere, directly and indirectly, through the
habits you've formed and the choices you make. How you heat and cool your
home, what kind of car you drive and how you maintain it, how hot a shower
you take, what kind of food you put on the table, and how you dispose of organic and inorganic waste—all these decisions make up your personal carbon
footprint.

You may already know many of the ways to reduce your contribution of carbon dioxide: turning off lights, turning down the thermostat, getting an "energy audit" of your home. But you may have put off making more daunting
changes. You may not have felt you could do much to change your daily car
commute or how often you have to fly for business.

True, you may not be able to afford to buy a hybrid car tomorrow, but arrange
to carpool, take the train, or telecommute even two days a week. If you must
drive, maintaining your car properly will improve gas mileage and reduce the
amount of carbon dioxide your car produces. You may not be able to avoid flying for business, but you can purchase carbon offsets for your flights.

Concentrate on doing what you can: put up a clothesline, lower the setting
on your water heater, and place a composter where it's easy to use and put a
note on the trash can to remind yourself to use it. At the store, look for and purchase products made from recycled materials when you have the option.

Remember, it takes time to change habits. You don't have to do it alone: make the changes a family project or a friendly competition among your co-workers or neighbors.

A Better Way to Travel

In our globe-trotting era, our travel choices have a profound effect on the planet. Whether you will be traveling for business or pleasure, make your plans with your carbon footprint in mind.

Choose bus or rail travel over air when you can. Mass transit is not only better for the environment, but also allows you to see more of the region you're visiting and allows you the time to catch up on sleep, reading, or work. If you must fly, consider an airline with a newer, fuel-efficient fleet (rankings are available at www.greenopia.com). Then make your trip climate neutral by making a donation to a zero- or negative-emission project, such as building renewable energy facilities or planting trees. Such so-called carbon offsets can now be added to your ticket with a single click on many online travel websites.

When choosing where to stay, lower your impact by staying in a hostel, arranging a timeshare—even camping, if practical! If you do book a hotel, look for one that is committed to reducing its impact on the environment. Some hotel chains are cutting down on the waste they send to landfills by installing soap and shampoo dispensers and making shower caps and sewing kits available only upon request. Once you arrive, arrange to have your linens and towels changed only upon request to save energy and water. Leave positive feedback for the hotel management, letting them know that their green policies were a major factor in giving the hotel your business.

To get around in a foreign city, forgo a rental car or taxi and make use of the local bus or rail system. Many cities are pedestrian- and bike-friendly; research walking tours and bike rentals before your departure. Walking and cycling are not only better for the environment and your health, but also allow you to discover hidden surprises you might otherwise overlook while speeding by on the highway. If you must rent a car during your trip, choose the most fuel-efficient model available; you may be able to choose a hybrid or electric car. Another option that reduces greenhouse gas emissions is car sharing, which reduces the number of cars on the road. ZipCar, the world's largest car-sharing network, runs car shares in the United States, Canada, Spain, and the United Kingdom and plans to expand in Europe.

Finally, be sure that any sightseeing tours you book are designed to minimize your impact on the local people and landscape of your destination. You may be familiar with the concept of ecotourism—travel that seeks to minimize its impact on the environment. More recently a newer "geotourism" movement has begun to emerge. Championed by the Center for Sustainable Destinations (CSD), geotourism seeks to "sustain or enhance" the geographical character of a place—not just its environment, but the culture, heritage, and aesthetics that make that destination unique. Geotourism also aims to enhance the well-being of the people who call that destination home.

If You Love the Ocean, Pass One On

"Live like you love the ocean." That is the simple message of sea turtle researcher and conservationist Wallace J. Nichols. In 2009 Nichols launched the Blue Marble campaign to raise ocean awareness in the year leading up to World Oceans Day 2010. Blue marbles symbolize the ocean as well as our water planet, which resembles a blue marble from space. The act of receiving a blue marble and passing it on spreads the message of ocean conservation, but it has the potential to be more—a catalyst for real change.

Nichols was moved to start the Blue Marble campaign by the fact that 95 percent of conservation efforts are aimed at land, even though the ocean covers three-quarters of Earth's surface. He wondered if he could use a viral campaign, powered by social media, to bring attention to the ocean's plight during 2010—the year of the first World Oceans Day recognized by the United Nations and the centenary of pioneer ocean conservationist Jacques Cousteau.

During that year sixty thousand blue glass marbles changed hands around the world. With each exchange, giver and receiver shared what they were doing to help the ocean and pledged to pass the marble, and their commitment, on. The images of the marbles' travels and their stories were posted and shared on blogs and websites, Facebook and Twitter.

The Blue Marble campaign didn't stop with World Oceans Day 2010. Little blue marbles are still circling our big blue marble, connecting people of all ages and all backgrounds, from all walks of life, in dozens of countries. The blue marbles are spreading the message that the ocean is in crisis and the solution lies in our hands.

Living Blue Resources

The list of resources and organizations below does not aim to be comprehensive, but rather to provide a starting point for locating authoritative information and advice for living blue and acting to save the ocean.

National Geographic Society's Ocean Initiative (including the global partnership Mission Blue)

www.ocean.nationalgeographic.com/ocean/

New England Aquarium's Live Blue Initiative

www.liveblueinitiative.org

Conservation International

www.conservation.org/global/marine

The Blue Frontier's Fifty Ways to Save the Ocean

www.bluefront.org

While not dedicated to ocean conservation, Appropedia's Green Wiki has a good guide to various green "wikis" with collaborative content, some in languages other than English.

www.appropedia.org/Green_wiki

Ecotourism and Green Travel

Lindblad Expeditions and Cruises, headquartered in New York City, partners with the National Geographic Society to provide sustainable travel options to world-class destinations.

www.expeditions.com

The Green Hotels Association lists member hotels in all fifty states and in many other countries.

www.greenhotels.com

Energy Efficiency, Home and Auto

The U.S. Environmental Protection Agency's Energy Star program has information on energy-efficient appliances, appliance rebate programs, home energy audits, and more.

www.energystar.gov

The American Council for an Energy-Efficient Economy has information on how to make your personal electronics—computers, printers, televisions, cameras, and almost anything else that plugs in—more efficient.

www.aceee.org

Here are some other sites that can provide more information on energy efficiency measures:

www.oee.nrcan.gc.ca (Canada)
www.uk-energy-saving.com (United Kingdom)
www.energyrating.gov.au (Australia)
www.Greencar.com
www.Planettran.com
www.driveclean.ca.gov (Canada)
www.greencarsite.co.uk (United Kingdom)
www.greenvehicleguide.gov.au (Australia)

Exotic Species and Marine Invasives

www.invasive.org (North America)
www.introduced-species.co.uk (United Kingdom)
www.environment.gov.au (Australia)
www.protectyourwaters.net (United States)

Green Landscaping

The Ecological Landscaping Association has lots of resources for ecologically minded landscaping for your yard and garden.

www.ecolandscaping.org

Pollinator Partnership has tips on inviting pollinators to your garden by planting regionally appropriate plants.

www.pollinator.org (North America)

The Lady Bird Johnson Wildflower Center has a native plant database and links to nurseries that supply native plants.

www.wildflower.org (North America)

The Graywater Policy and Science Center has information on capturing wastewater from the laundry, kitchen, and shower for use outdoors.

www.graywater.org

Home Aquariums

Reef Protection International has an online Reef Fish Guide for the responsible home aquarium hobbyist.

http://reefprotect.org/fish_guide.htm

Local and Organic Food

Here are some other sites that can provide more information on local and healthy eating:

Local Harvest helps connect you to farmers' markets and community-sponsored agriculture (CSAs) throughout the United States.

www.localharvest.org
www. cog.ca (Canadian Organic Growers)
www.eatwellguide.org
www.farmshopping.net (United Kingdom)
www.organicfooddirectory.com.au (Australia)

Ocean Advocacy and Political Action

Conservation International (CI) empowers societies to responsibly and sustainably care for nature, our global biodiversity, for the well-being of humanity. With headquarters in Washington, DC, CI works in more than forty countries on four continents.

www.conservation.org

International Union for Conservation of Nature (IUCN), the world's oldest and largest global environmental network, helps the world find pragmatic solutions to our most pressing environmental and development challenges. The mission of its Protect Planet Ocean is to inspire and inform better protection of our ocean through the establishment of a global representative network of successfully managed marine protected areas (MPAs).

www.protectplanetocean.org

The League of Conservation Voters provides a good home base for conservation-minded activism, from scorecards for elected officials to current ballot initiatives and legislation in the United States. Sign up for their RSS feed to keep abreast of environmental issues without flooding your inbox with multiple updates.

www.lcv.org

Since 1951 The Nature Conservancy has organized and led conservation efforts around the world, working to protect ecologically important lands and waters for nature and people.

www.nature.org

Founded in 1972, Ocean Conservancy is an environmental nonprofit advocacy organization that promotes healthy and diverse ocean ecosystems and opposes practices that threaten marine life and human life.

www.oceanconservancy.org

The Sylvia Earle Alliance is committed to creating and protecting Hope Spots, special places that are critical to the health of the ocean.

www.sylviaearlealliance.org

The World Wildlife Fund is a global conservation organization acting locally through a network of over ninety offices in over forty countries around the world.

www.wwf.org

The following organizations work to protect the world's coral reefs:

http://reefcheck.org/
http://reefprotect.org/
www.coral.org

Recycling

www.recycle.net (United States)
http://ec.gc.ca/ (Canada)
http://recyclenow.com/ (United Kingdom)
www.recycleaustralia.org (Australia)

The Plastic Pollution Coalition website walks you through the steps to kick the plastic habit. You also can follow the Plastics Pollution Coalition on Facebook and Twitter.

http://plasticpollutioncoalition.org/

Sustainable Seafood

The New England Aquarium features updated information on its website on ocean-friendly choices, including recipes for the home kitchen.

www.neaq.org/sustainableseafood

FishChoice is a portal to connect commercial seafood buyers with sustainable seafood suppliers.

www.fishchoice.com

Green Chefs, Blue Ocean, a partnership between Blue Ocean Institute and Chefs' Collaborative, includes an interactive online curriculum about seafood sustainability for chefs and culinary students.

www.oceanfriendlychefs.org

Marine Stewardship Council is a global seafood certification program under which wild-capture fisheries can become certified as sustainable, environmentally responsible, and well managed. Look for the blue eco-label in stores and restaurants.

www.msc.org

Monterey Bay Aquarium's Seafood Watch has a guide to sustainable seafood that you can download to your mobile phone.

www.seafoodwatch.org

SeaChoice, Canada's most comprehensive sustainable seafood program, is led by five internationally respected Canadian conservation organizations.

www.seachoice.org

Seafood Choices Alliance has an Afishianado e-newsletter to keep you up to date on current industry and consumer trends, new market research, and sustainable seafood efforts.

www.seafoodchoices.org

Volunteer Opportunities, Citizen Science, and Ocean Education

NOAA has opportunities for volunteers in U.S. National Marine Sanctuaries.

www.volunteer.noaa.gov

Blueventures offers opportunities to participate in marine conservation projects around the world.

www.blueventures.org

If you are a parent, teacher, or mentor of a student, consider coaching a team for the National Ocean Sciences Bowl competition, sponsored by the Consortium for Ocean Leadership.

www.nosb.org

Find citizen science opportunities in your own backyard or farther afield at this clearinghouse, which helps match interested citizen scientists with projects. Sample projects include Jellyfish Watch and EarthDive, a global project recruiting recreational divers to monitor key indicator species and log data using Google Earth.

www.scienceforcitizens.net

You can also participate in a BioBlitz, a 24-hour event in which ordinary people of all ages work alongside scientists to find and identify as many species of plants, animals, microbes, fungi, and other organisms as possible. Find National Geographic BioBlitz on Facebook and at

www.nationalgeographic.com.

For a collection of marine education resources, visit the Bridge, a website supported by the National Sea Grant Office, the National Oceanographic Partnership Program (NOPP), and the National Marine Educators Association (NMEA).

www.vims.edu/bridge

In addition to the opportunities listed immediately above, many of the organizations already mentioned offer opportunities for volunteers to get involved at the local and global scales.

PHOENIX ISLANDS TIMELINE

70 mya   The Tokelau Seamount Chain, including the Phoenix Islands, is formed, emerging from the ocean first as volcanoes, then subsiding and forming into reefs over millions of years. The islands are among the world's oldest formations produced by living organisms.

Pre-1500   Early Polynesian and Micronesian voyagers reach the Phoenix Islands. While they left archaeological evidence of visits and temporary settlements, the arid conditions on the islands prevented permanent settlement.

1832   Cartographer Andrew Goldsmith shows Sydney and Birnie Islands on his map of the Pacific Ocean.

1838   The U.S. Exploring Expedition (the "Wilkes" expedition of 1838–42) collects detailed information on 280 Pacific islands, including the exact positions of several of the Phoenix Islands.

1842   Charles Darwin mentions Phoenix Island (Rawaki) and Sydney Island (Manra) in his treatise on the origin of coral reefs.

1854   The New Bedford whaler *Canton* runs aground on the island that now bears its name. The Phoenix Islands lay "on the Line," the favored whaling grounds for hunting of the sperm whale, a species decimated by 1900.

1856   The U.S. Congress passes the Guano Islands Act of 1856; in ensuing decades, the United States lays claim to Enderbury, Phoenix, Sydney, Hull (Orona), and Gardner (Nikumaroro) Islands.

1937   Britain includes the Phoenix group in the Gilbert and Ellice Islands, a crown colony. This was the last attempted colonial expansion of the British Empire (the Phoenix Islands Settlement Scheme). On July 2, aviator Amelia Earhart and navigator Fred Noonan disappear in the vicinity of the Phoenix Islands. In July astronomers from New Zealand observe a total eclipse of the sun from Canton (Kanton) Island.

1938–39   The United States claims sovereignty over Canton and Enderbury in 1938. In 1939 Britain and the United States agree to exercise joint control over the two islands for fifty years as the Canton and Enderbury Islands condominium. Colonial administrator Harry Maude resettles sixty-one settlers from the overcrowded Gilbert Islands on Sydney, Hull, and Gardner. The population peaks at 1,300 in the mid-1950s, but after a drought, Britain evacuates settlers to the Solomon Islands.

1940–41   Pan American Airlines runs its Pacific Clipper Ship service, with a refueling stop at Canton, until the outbreak of World War II.

1957 Birnie, Canton, Enderbury, and Hull are identified as "prohibited areas" under the Prohibited Areas Ordinance, effectively prohibiting access to the islands, though without specific management measures or goals.

1962 U.S. astronaut John Glenn sees the sun rise over Canton Island as he orbits the Earth in *Friendship 7*. NASA keeps a listening post on Canton through much of the Cold War.

1975 Birnie, Phoenix, and McKean Islands are declared wildlife sanctuaries under the Wildlife Conservation Ordinance.

1979 The Line, Gilbert, and Phoenix Islands achieve independence from Britain and the United States as the new Republic of Kiribati.

1979 Kiribati and the United States sign a Treaty of Friendship in which they agree to cooperate in the conservation and management of the Phoenix and Line Islands.

1983 Martin Garnett prepares the first management plan for the Phoenix Islands focusing on the islands and their resources.

1989 The International Group for Historic Aircraft Recovery (TIGHAR) stages the first of many expeditions to Nikumaroro in search of Earhart's plane.

1999 TIGHAR sends its fifth expedition party to Nikumaroro, the group's second charter of *Nai'a*. Rob Barrel and Cat Holloway, owners of Nai'a Cruises, approach Greg Stone to report on the spectacular, unspoiled reefs there.

2000 New England Aquarium (NEAq) sponsors the first expedition to the Phoenix Islands, led by Stone.

2001 Stone, David Obura, and Sangeeta Mangubhai meet with Kiribati officials to broach the idea of a marine protected area (MPA); meanwhile, a shark-finning boat spends three months in the islands, stripping sharks from four of the eight islands. Kiribati initiates a new program to resettle residents from the Gilberts on Orona. The so-called "Kakai Scheme," which aimed to employ the settlers in the harvest of copra, sea cucumbers, and shark fins, was unsuccessful, and the Orona colony was disbanded in 2004.

2002 Stone and Obura lead a second NEAq expedition to the Phoenix Islands. Early signs of coral bleaching are observed.

2002–3 A severe El Niño event brings sustained high ocean temperatures to the Pacific; corals respond by bleaching.

2005 Kiribati, NEAq, and Conservation International (CI) sign an agreement to design and establish an MPA in the Phoenix Islands.

2006 The government of Kiribati declares the Phoenix Islands Protected Area (PIPA). Dr. Ray Pierce leads an expedition to survey the islands for invasive species and create a plan for restoring the bird populations to historical levels.

2007 Roger Uwate and Tukabu Teroroko undertake the task of valuing the natural resources of the Phoenix Islands, compiling and analyzing more than seven hundred reports and scientific papers. Kiribati implements interim management measures agreed to while the process of formally establishing PIPA proceeds.

2008 PIPA is legally established under Kiribati law, with expanded boundaries making it, at the time, the largest MPA in the world. Management planning and the full operation of the PIPA oversight body, the PIPA Management Committee, begins. Ray Pierce leads island restoration and invasive species eradication, starting with Rawaki and McKean.

January 2009 The United States declares Howland and Baker Islands (geographically part of the Phoenix Islands) part of the Pacific Remote Islands Marine National Monument (PRIMNM), with protection extending out to 50 nautical miles around each island. Later that month, Kiribati nominates PIPA for listing as a UNESCO World Heritage site. A trawler illegally fishing in the PIPA is detained under a United States–Kiribati Shipriders Agreement and fined A$4.8 million.

March 2009 The United Nations Global Environment Facility approves US$1 million for PIPA support.

May 2009 Greg Stone becomes Senior Vice President and Chief Scientist for Oceans at Conservation International; he remains Senior Vice President of Exploration and Conservation at NEAq. Kiribati passes PIPA Conservation Trust legislation.

September 2009 NEAq leads a fourth expedition. A sister-site agreement is signed with the Papahānaumokuākea Marine National Monument (PMNM) in Hawaii, a first for large MPAs globally. The sister MPAs aim to learn from each other and cooperate in research and management of their sites. Kiribati's cabinet approves the PIPA 2010–2014 Management Plan.

November 2009 Ray Pierce heads back to the islands of McKean and Rawaki to find that the first invasive species eradications have been successful and the islands' wildlife is flourishing in the absence of the introduced pests.

March 2010 The PIPA Conservation Trust holds its first meeting in Tarawa, Kiribati.

April 2010 Britain announces the establishment of the Chagos Archipelago Reserve in the Indian Ocean, making PIPA the world's second largest MPA. PIPA remains the largest MPA in the Pacific Ocean and the largest ever committed to by a developing nation.

August 2010 The PIPA is successfully inscribed on the list of UNESCO World Heritage sites, making it the world's largest World Heritage site.

September 2010 Scientists meet in Boston for the first Phoenix Islands Scientific Research Agenda Meeting to set research priorities for the next ten years.

July 2011 Ray Pierce leads another mission to PIPA as part of the first ever three-country (United Kingdom, United States, Kiribati) collaborative pest-eradication expedition. Rats are targeted on Enderbury and Birnie.

August 2011 The PIPA Conservation Trust hires its first executive director.

2012 PIPA's success and leverage continues.

APPENDIX

NEW ENGLAND AQUARIUM RESEARCH EXPEDITIONS TO THE PHOENIX ISLANDS

Expedition 1: June 24–July 15, 2000

Nikumaroro (4 days), McKean (1 day), Kanton (3 days), Enderbury, Rawaki (2 days), Manra (1 day), Orona (2 days)

SCIENTIFIC TEAM

Gregory S. Stone, Ph.D., New England Aquarium (NEAq)

Austen Yoshinaga-Stone, NEAq

Steven L. Bailey, Ichthyologist, Coral Reef Assessments, NEAq

David Obura, Ph.D., Coastal Oceans Research and Development in the Indian Ocean (CORDIO), East Africa

Sangeeta Mangubhai, World Wildlife Fund (WWF) South Pacific

NAI'A CRUISES

Robert Barrel, Captain

Cat Holloway, Divemaster

ADDITIONAL PERSONNEL

Tuake Teema, Kiribati Department of Fisheries

Alex Morrison, Camera Operator, Topside

PASSENGER-SPONSORS

Bruce Thayer

Mary Jane Adams, M.D.

Craig Cook, M.D.

Expedition 2: June 5–July 10, 2002

Nikumaroro (8 days), Manra (2 days), Rawaki (1 day), Kanton (9 days), Enderbury (2 days), Birnie (1 day), Orona (4 days)

SCIENTIFIC TEAM

Greg Stone

Austen Yoshinaga-Stone

David Obura

Sangeeta Mangubhai

Gerald R. Allen, Ph.D. Senior Ichthyologist, Conservation International (CI)

Steven L. Bailey

Alistair Hutt, New Zealand Department of Conservation

NAI'A CRUISES

Robert Barrel

Cat Holloway

ADDITIONAL PERSONNEL

Paul Nielson, Ministry of Environment, Lands, and Agricultural Development (MELAD)

Paul Nicklen, National Geographic Society (NGS)

Joe Stancampiano, NGS

PASSENGER-SPONSOR

Mary Jane Adams

Expedition 3: May 15–27, 2005

Enderbury (1 day), Kanton (3 days), Manra (1 day), Nikumaroro (4 days), Orona (3 days), Rawaki (1 day)

SCIENTIFIC TEAM

David Obura

Sangeeta Mangubhai

NAI'A CRUISES

Robert Barrel

Expedition 4: September 13–23, 2009

Nikumaroro (2 days), McKean (1 day), Kanton (3 days), Enderbury (1 day), Rawaki (1 day), Orona (2 days)

SCIENTIFIC TEAM

Greg Stone

David Obura

Randi Rotjan, Ph.D., NEAq

Les Kaufman, Ph.D., Boston University and CI

Stuart Sandin, Ph.D., Scripps Institution of
   Oceanography (SIO)

Kate Madin, Woods Hole Oceanographic
   Institution (WHOI)

Larry Madin, Ph.D., WHOI

NAI'A CRUISES

Robert Barrel

ADDITIONAL PERSONNEL

Tukabu Teroroko, Director, Phoenix Islands
   Protected Area

Tuake Teema

Jeff Wildermuth, NGS

Brian Skerry, NGS

PASSENGER-SPONSORS

Alan Dynner, NEAq Board of Overseers

Craig Cook

Jim Stringer

SPECIES TALLY

Published Records of Known Species in the Phoenix Islands*

Algae	107 (South et al. 2001; Obura et al. 2011a)
Arthropods	50 (Degener and Gillaspy 1955; Banner and Banner 1964)
Birds	20 (Buddle 1938; Kepler 2000)
Bivalves	43 (Bryan 1974)
Corals	> 120 (Obura and Stone 2002)
Dolphins and whales	20 (Reeves et al. 1999)
Fishes	516 (Allen and Bailey 2011)
Insects	93 (Van Zwaluwenburg 1955)
Other invertebrates (spiders, crabs, scorpions)	15 (Van Zwaluwenburg 1955)
Plants	28 native; 59 introduced (Fosberg and Stoddart 1994)
Reptiles	5 (Pierce et al. 2006; Obura et al. 2011b)

References

Allen, G., and S. Bailey. 2011. "Reef Fishes of the Phoenix Islands, Central Pacific Ocean." *Atoll Research Bulletin*, no. 589: 83–118.

Banner, A. H., and D. M. Banner. 1964. "Contributions to the Knowledge of the Alpheid Shrimp of the Pacific Ocean. 9. Collection from the Phoenix and Line Islands." *Pacific Science* 18, no. 1: 83–100.

Bryan, E. H., Jr. 1974. *Panala'au Memoirs*. Pacific Science Information Center, B. P. Bishop Museum.

Buddle, G. A. 1938. "Notes on the Birds of Canton Island." *Record of the Auckland Institute and Museum* 2, no. 3: 125–32.

Degener, O., and E. Gillaspy. 1955. "Canton Island, South Pacific." *Atoll Research Bulletin*, no. 41: 1–50.

Fosberg, F. R., and D. R. Stoddart. 1994. "Flora of the Phoenix Islands, Central Pacific." *Atoll Research Bulletin*, no. 393: 1–60.

Kepler, A. K. K. 2000. *Report: Millennium Sunrise, Line and Phoenix Islands Expedition, December 15, 1999–January 28, 2000*. Athens, GA: Pan-Pacific Ecological Consulting.

*These are only the published records. Actual numbers and taxa are assumed to be higher in this remote and still relatively unexplored region.

Obura, D., and G. Stone. 2002. *Islands: Summary of Marine and Terrestrial Assessments Conducted in the Republic of Kiribati, June 5–July 10, 2002*. Primal Ocean Project Technical Report NEAq-03-03. Boston: New England Aquarium.

Obura, D., G. Stone, S. Mangubhai, S. Bailey, A. Yoshinaga, C. Holloway, and R. Barrel. 2011a. "Baseline Marine Biological Surveys of the Phoenix Islands, July 2000." *Atoll Research Bulletin*, no. 589: 1–62.

———. 2011b. "Sea Turtles of the Phoenix Islands, 2000–2002." *Atoll Research Bulletin*, no. 589: 119–24.

Pierce, R. J., T. Etei, V. Kerr, E. Saul, A. Teatata, M. Thorsen, and G. Wragg. 2006. *Phoenix Islands Conservation Survey and Assessment of Restoration Feasibility: Kiribati*. Report prepared for Conservation International and Pacific Islands Initiative, Auckland University, New Zealand.

Reeves, R. R., S. Leatherwood, G. S. Stone, and L. G. Eldredge. 1999. *Marine Mammals in the Area Served by the South Pacific Regional Environment Programme (SPREP)*. South Pacific Regional Environment Programme (SPREP).

South, G. R., P. A. Skelton, and A. Yoshinaga. 2001. "Subtidal Benthic Marine Algae of the Phoenix Islands, Republic of Kiribati, Central Pacific." *Botanica Marina* 44, no. 6: 559–70.

Van Zwaluwenburg, R. H. 1955. "The Insects and Certain Other Arthropods of Canton Island." *Atoll Research Bulletin*, no. 42: 1–11.

BIBLIOGRAPHY

Chapter 1

Bester, Cathleen. "Bluntnose Sixgill Shark." Ichthyology at the Florida Museum of Natural History. Accessed November 11, 2010. http://www.flmnh.ufl.edu/fish/Gallery/Descript/Bsixgill/Bsixgill.html.

Encyclopedia of Life. "*Hexanchus griseus* (Bonnaterre 1788)." Accessed November 11, 2010. http://www.eol.org/pages/212027.

"Hawaii Says No to Shark Fin Soup." *Science News*, May 29, 2010. Accessed November 11, 2010. http://www.upi.com/Science_News/2010/05/29/Hawaii-says-no-to-shark-fin-soup/UPI-35471275161131/.

International Union for the Conservation of Nature (IUCN) Shark Specialist Group. 2003. *Shark Finning*. Paper presented at the IUCN, Malaga, Spain.

Chapter 2

Anderson, A., P. Wallin, B. Frankhauser, and G. Hope. 2000. "Toward a First Prehistory of Kiritimati (Christmas) Island, Republic of Kiribati." *Journal of the Polynesian Society* 109, no. 3: 273–93.

di Piazza, A., and Erik Pearthree. 2004. *Sailing Routes of Old Polynesia: The Prehistoric Discovery, Settlement and Abandonment of the Phoenix Islands*. Vol. 11. Bishop Museum Bulletins. Honolulu: Bishop Museum Press.

Lewis, David. 1994. *We, the Navigators: The Ancient Art of Landfinding in the Pacific*. 2nd ed. Honolulu: University of Hawaii Press.

Skaggs, Jimmy. 1995. *The Great Guano Rush: Entrepreneurs and American Overseas Expansion*. New York: St. Martin's Press.

Chapter 3

Allen, Gerry, and Steven Bailey. 2011. "Reef Fishes of the Phoenix Islands, Central Pacific Ocean." *Atoll Research Bulletin*, no. 589: 83–118.

Kaufman, Les. 2005. "One Fish, Two Fish, Red Fish, Blue Fish: Why Are Coral Reefs So Colorful?" *National Geographic*, May.

Chapter 4

Burger, Alan E. 2005. "Dispersal and Germination of Seeds of *Pisonia grandis*, an Indo-Pacific Tropical Tree Associated with Insular Seabird Colonies." *Journal of Tropical Ecology* 21, no. 3: 263–71. Accessed January 26, 2011. www.natureseychelles.org.

Clapp, R. B., and F. C. Sibley. 1967. "New Records of Birds from the Phoenix and Line Islands." *Ibis* 109: 122–25.

Davis, N. E., D. J. O'Dowd, R. Mac Nally, and P. T. Green. 2009. "Invasive Ants Disrupt Frugivory by Endemic Island Birds." *Biology Letters* 6, no. 1: 85–88. doi:10.1098/rsbl.2009.0655.

Garnett, M. C. 1983. "A Management Plan for Nature Conservation in the Line and Phoenix Islands. Part 1. Description." Unpublished report prepared for the Ministry of Line and Phoenix Islands, Kiritimati Island, Kiribati.

Pierce, R. J., N. Anterea, U. Anterea, K. Broome, D. Brown, L. Cooper, H. Edmonds, F. Muckle, W. Nagle, G. Oakes, M. Thorsen, and G. Wragg. 2008. *Operational Work Undertaken to Eradicate Mammalian Pests in the Phoenix Islands, Kiribati, May–June 2008*. Eco Oceania Ltd. report for Government of Kiribati and NZAID.

Pierce, R. J., T. Etei, V. Kerr, E. Saul, A. Teatata, M. Thorsen, and G. Wragg. 2006. *Phoenix Islands Conservation Survey and Assessment of Restoration Feasibility: Kiribati*. Report prepared for Conservation International and Pacific Islands Initiative, Auckland University, New Zealand.

Sibley, F. C., and R. B. Clapp. 1967. "Distribution and Dispersal of Central Pacific Lesser Frigatebirds *Fregata ariel*." *Ibis* 109: 328–37.

Steadman, David. 2006. *Extinction and Biogeography of Tropical Pacific Birds*. Chicago: University of Chicago Press.

Uwate, K. R., W. Fitzgerald, and T. Teroroko. 2007. *Valuation of Phoenix Islands Marine and Fisheries Resources*. Phoenix Islands Protected Area, Ministry of Environment, Lands, and Agricultural Development, Government of Kiribati.

Veitch, C. R., and M. N. Clout, eds. 2002. *Turning the Tide: The Eradication of Invasive Species*. Proceedings of the International Conference on Eradication of Island Invasives. Occasional Paper of the IUCN Species Survival Commission, no. 27. Cambridge, UK: IUCN.

Chapter 5

Gunderson, Ross. 1997. "Through a Looking Glass: Micromollusks." *American Conchologist*, March. Accessed November 14, 2010. http://www.conchologistsofamerica.org/articles/y1997/9703_gunderson.asp.

Chapter 6

European Project on Ocean Acidification. "Acidification: The Facts." Accessed October 26, 2010. http://www.epoca-project.eu/index.php/what-is-ocean-acidification/faq.html.

Obura, D. 2011. "Coral Reef Structure and Zonation of the Phoenix Islands." *Atoll Research Bulletin*, no. 589: 63–82.

Obura, D., and S. Mangubhai. 2011. "Coral Mortality Associated with Thermal Fluctuations in the Phoenix Islands, 2002–2005." *Coral Reefs* 30, no. 3: 607–19.

Stone, G. 2011. "Phoenix Rising." *National Geographic* 219, no. 1: 70–84.

———. 2004. "Phoenix Islands: South Pacific Hideaway." *National Geographic* 205, no. 2: 48–65.

Stone, G., D. Obura, and R. Rotjan. 2009. *Phoenix Islands Protected Area Assessment and Expedition Report*. Boston: New England Aquarium.

Chapter 7

The full quote from Attenborough reads: "Tourism is a mixed blessing for the Galapagos but the fact is, if there was not tourism to the islands and the local people did not get any income from it, there would be nothing left there now. It would all be gone. It is the lesson of conservation around the world that unless the people who live in such places, whose land they feel belongs to them, are on the side of conservation, you're doomed. So tourism, if it's evil, is a necessary evil and one that in this instance can be controlled." The essay originally appeared in *Lonely Planet* magazine and was quoted on the Ecotourism Society of the Seychelles blog (accessed October 30, 2010), http://ecotourismseychelles.blogspot.com.

Food and Agriculture Organization of the United Nations. "Kiribati National Fishery Sector Overview, Aid." Accessed January 20, 2011. http://www.fao.org/fisher/countrysector /FI-CP KI/en.

Jupiter, Stacy D., and Daniel P. Egli. 2011. "Ecosystem-Based Management in Fiji: Successes and Challenges after Five Years of Implementation." *Journal of Marine Biology*, 1–14. Accessed January 5, 2012. doi:10.1155/2011/940765.

Sanders, Jane M. 2005. "Reefing the Benefits." *GT (Georgia Tech) Research Horizons* (Fall): 1–4.

Thomas, Frank R. 2003. "'Taming the Lagoon': Aquaculture Development and the Future of Customary Marine Tenure in Kiribati, Central Pacific." Special issue, *Geografiska Annaler*, Series B, Human Geography 85, no. 4: 243–52.

United Nations Economic and Social Commission for Asia and the Pacific. 2003. "Kiribati." In *ESCAP Tourism Review No. 23: Ecotourism Development in the Pacific Islands*, 16–20. New York: UNESCAP.

Chapter 8

Uwate, K. R., W. Fitzgerald, and T. Teroroko. 2004. *Valuation of Phoenix Islands Marine and Fisheries Resources*. Phoenix Islands Protected Area, Ministry of Environment, Lands and Agricultural Development, Government of Kiribati.

Chapter 9

Jupiter, Stacy D., and Daniel P. Egli. 2011. "Ecosystem-Based Management in Fiji: Successes and Challenges after Five Years of Implementation." *Journal of Marine Biology*, 1–14. Accessed January 5, 2012. doi:10.1155/2011/940765.

Nomination for a World Heritage Site. 2009. Phoenix Islands Protected Area, Kiribati. Tarawa, Kiribati: Phoenix Islands Protected Area (PIPA) Office.

Rotjan, R. D., and D. O. Obura. 2010. *Phoenix Islands Protected Area Ten-Year Research Vision*. New England Aquarium Reports.

Chapter 10

Amos, Amy Mathews, and John D. Claussen. 2009. *Certification as a Conservation Tool in the Marine Aquarium Trade: Challenges to Effectiveness*. Final Report, May 2009. Turnstone Consulting/Starling Resources.

Fisher, E. 2010. "Ocean Hero Finalists: The Shark Finatics." *The Beacon*, May 12. http:// na.oceana.org/en/blog/2010/05/ocean-hero-finalists-the-shark-finatics.

"Invasive Species: Stop the Alien Attack." The National Audubon Society. Accessed January 26, 2012. http://www.stopinvasives.org.

FURTHER READING

Burke, Loretta, and Jon Maidens. 2004. "Reefs at Risk in the Caribbean." World Resources Institute. Accessed January 21, 2011. http://www.wri.org/publication/reefs-risk-caribbean.

Cramer, Deborah. 2008. *Smithsonian Ocean: Our Water, Our World*. New York: Smithsonian Books.

Ellis, Richard. 2003. *The Empty Ocean: Plundering the World's Marine Life*. Washington, DC: Island Press.

Flannery, Tim. 2006. *The Weather Makers: The History and Future Impact of Climate Change*. New York: Atlantic Monthly Press.

Gaston, Anthony A. 2004. *Seabirds: A Natural History*. New Haven, CT: Yale University Press.

Greenberg, Paul. 2010. *Four Fish: The Future of the Last Wild Food*. New York: Penguin.

Helvarg, David. 2006. *50 Ways to Save the Ocean*. Makawao, HI: Inner Ocean Publishing.

Humann, Paul. 2002. *Reef Set: Reef Fish, Reef Creature, and Reef Coral*. 3 vols. Jacksonville, FL: New World Publications.

Lal, Brij, and Kate Fortune. 2000. *The Pacific Islands: An Encyclopedia*. Honolulu: University of Hawaii Press.

Last, P. 2009. *Sharks and Rays of Australia*. 2nd ed. Cambridge, MA: Harvard University Press.

National Wildlife Federation. 2010. *Ecotourists Save the World: The Environmental Volunteer's Guide to More Than 300 International Adventures to Conserve, Preserve, and Rehabilitate Wildlife and Habitats*. New York: Perigee.

Pepperell, Julian. 2010. *Fishes of the Open Ocean: A Natural History and Illustrated Guide*. Chicago: University of Chicago Press.

Philbrick, Nathaniel. 2005. *Sea of Glory: The Epic South Seas Expedition, 1838–1842*. London: Harper Perennial.

Randall, John E. 2005. *Reef and Shore Fishes of the South Pacific: New Caledonia to Tahiti and the Pitcairn Islands*. Honolulu: University of Hawaii Press.

Scubazoo. *Reef*. 2007. 1st ed. New York: Dorling Kindersley.

Trautman, James. 2007. *Pan American Clippers: The Golden Age of Flying Boats*. Erin, ON: Boston Mills Press.

Tsiao, Sunny. 2008. *"Read you loud and clear!" The Story of NASA's Spaceflight Tracking and Data Network*. Washington, DC: National Aeronautics and Space Administration, History Division, Office of External Relations.

Veitch, C. R., and M. N. Clout, eds. 2002. *Turning the Tide: The Eradication of Invasive Species*. Proceedings of the International Conference on Eradication of Island Invasives. Occasional Paper of the IUCN Species Survival Commission, no. 27. Cambridge, UK: IUCN.

Wendt, Albert, ed. 1995. *Nuanua: Pacific Writing in English Since 1980*. Honolulu: University of Hawaii Press.

Wilson, Edward. 2010. *The Diversity of Life*. Questions of Science. New ed. Cambridge, MA: Belknap Press of Harvard University Press.

ACKNOWLEDGMENTS

With a body of work this vast, and over so many years, it is impossible to thank all the many wonderful people, countries, and prestigious organizations that have contributed. But we will try. So our apologies to any who are not mentioned; we thank you as well.

To begin, we thank the resilient and remarkable people of the largest atoll nation in the world, Kiribati; for they, through their hard work and enthusiastic support of this initiative, have made the Phoenix Islands Protected Area (PIPA) possible. Of course, people are inspired, informed, and led by great leaders, and in this category Kiribati is well endowed. We have been so impressed by the wisdom, thoughtfulness, and courage shown by these leaders throughout the development of PIPA. Te Beretitenti His Excellency President Anote Tong is the most inspiring leader we have ever met. It was his vision that made PIPA possible. Other elected and government officials instrumental in the development of PIPA, particularly in its early years, include Tetabo Nakara, Martin Puta Tofinga, Kaburoro Ruaia, and Amberoti Nikora. We also acknowledge the commitment of the officials and individuals in Kiribati who have nurtured PIPA and still serve in government and the PIPA office and Management Committee today—theirs is the responsibility of maintaining this jewel in the Pacific.

Special thanks go to Mr. Tukabu Teroroko, the first director of PIPA, and to Dr. Teuea Toatu, the first executive director of the PIPA Trust, and to Cat Holloway and Rob Barrell for pioneering the access and originally taking us to the Phoenix Islands.

The work and commitment of many scientists and naturalists who have visited the Phoenix Islands have also enabled this project to happen. The contributors to this book all show their passion, knowledge, and commitment, and we would like to add a few names of those who have been part of this from the beginning: Sangeeta Mangubhai, Kandy Kendall, Bruce Thayer, and Craig Cook. To the organizations and donors that have showed uncommon commitment to PIPA, we would like to recognize NAI'A Cruises, the New England Aquarium, Conservation International, the Global Conservation Fund, the National Geographic Society, the Oak Foundation, the Pew Fellowship in Marine Conservation, the Critical Ecosystem Partnership Fund, the governments of New Zealand and Australia, the Akiko Shiraki Dynner Fund for Ocean Exploration and Conservation, the David and Lucile Packard Foundation, the United Nations Environment Programme, the Global Environment Facility, and Mares Dacor. We thank an anonymous donor who funded one of our expeditions to the islands and helped with the publication of this book.

Thanks go to Alan Dynner, Peter Seligmann, Russ Mittermeier, Kathy Moran, John Francis, Christian Parker, Tessie Lambourne, David Lambourne, Bud Ris, Chris Stone, Steve Katona, Sylvia Earle, Romas Garbaliauskas, Sebastian Troeng, Tarsu Murdoch, Jim Maragos, Christine Greene, and Bill and Barbara Burgess. In addition to the authors who contributed photos, we especially thank the photographers who enabled us to so richly and powerfully illustrate this book, including Jim Stringer, contributor Cat Holloway, and two National Geographic photographers, Brian Skerry and Paul Nicklen. Without images, it is impossible to tell the story of this amazing place on our planet. We thank Regen Jamieson for countless hours compiling the information in this book, and Erin DeWitt, Christie Henry, and Amy Krynak from the University of Chicago Press.

We thank the United Nations, Educational, Scientific and Cultural Organization (UNESCO) World Heritage program for inscribing PIPA as the largest and deepest World Heritage site on earth.

I (Greg) thank my wife, Austen Yoshinaga Stone, for participating on two grueling expeditions to PIPA and for hundreds of days when I was absent, either at PIPA or in Tarawa, working on this project. Thank you Doe Coover, our agent, who spent years thinking about, discussing, and finding the perfect publisher for this work, the University of Chicago Press.

There is one person more than any other that brought the many pieces of this book together into the semblance of a whole, and that is the "editor of editors," Ann Downer-Hazell. You are amazing, Ann.

Finally, we thank the legions of scientists, policy makers, divers, the general public, and others who are part of the current sea change in the effort to create a global network of large Marine Protected Areas for the benefit of humanity, of which PIPA was one of the first.

ABOUT THE CONTRIBUTORS

Honorable Anote Tong, President of the Republic of Kiribati

The son of a Chinese father and Gilbertese mother, Anote Tong graduated from Canterbury University in Christchurch, New Zealand, in 1974 with a degree in science, then gained a Master of Science in Economics from the London School of Economics. Between 1983 and 1992, he directed the Atoll Research and Development Unit at the University of the South Pacific. He entered Kiribati politics in 1994, serving as Minister of Natural Resources Development from 1994 to 1996. In 1996 he won a seat in the Kiribati parliament as a member of the Boutokaan Te Koaua Party, in which he served until his election to the presidency in 2003. Tong was reelected to a second term as president on October 17, 2007, and to a third and final term in January 2012.

President Tong has attracted international attention by warning that his country may become uninhabitable as soon as 2050 due to rising sea levels and salinization brought on by climate change. He was presented with an Outstanding Conservation Leadership Award by Conservation International (CI) in March 2006. For his global leadership in the field of marine conservation, the New England Aquarium presented President Tong with the David B. Stone Medal on the occasion of the declaration of the Phoenix Islands Protected Area (PIPA).

The Editors

GREGORY S. STONE, PH.D., is Executive Vice President and Chief Scientist for Oceans with Conservation International and Senior Vice President for Exploration and Conservation at the New England Aquarium. Stone is an ocean scientist who has written prolifically for scientific and popular publications, including *Nature* and *National Geographic*. He has delivered lectures throughout the world, including a 2010 TED talk aboard a ship in the Galápagos Islands and an address to the World Economic Forum in Davos, Switzerland. He has produced an award-winning series of marine conservation films. His book *Ice Island*, about Antarctica, won the 2003 National Outdoor Book Award for Nature and the Environment. A specialist in undersea technology and exploration, he has piloted deep-sea submersibles, lived in undersea habitats, and logged thousands of hours scuba diving in all the oceans of the world. He is a National Fellow of the Explorers Club, a recipient of the Pew Fellowship for Marine Conservation, and was awarded the National Science Foundation/U.S. Navy Antarctic Service medal for his research in Antarctica. Stone is chair of the World Economic Forum's Global Agenda Council on Ocean Governance. He now makes his home on the Big Island of Hawaii.

DAVID OBURA, PH.D., is founding director of CORDIO (Coastal Oceans Research and Development in the Indian Ocean) East Africa and an adjunct senior scientist at the New England Aquarium. Obura received his Ph.D. from the University of Miami in 1995 with a dissertation on coral bleaching and life history strategies, which has developed into a primary research interest in climate change, coral bleaching, and the resilience of coral reefs. He also helps artisanal fishers in East Africa to develop monitoring techniques and research tools, and he conducts remote-reef surveys in the central Pacific and the Indian Ocean. Obura chairs IUCN's Coral Specialist Group and the Climate Change and Coral Reefs Working Group. He lives in Mombasa, Kenya.

The Contributors

MARY JANE ADAMS, M.D., began diving in 1975 and immediately became addicted. Underwater photography soon became an important part of her diving, and she has amassed a large collection of marine life images. She enjoys studying the scientific aspects of the fishes and invertebrates captured in her photos. A dedicated citizen scientist, she participates in three or four liveaboard dive trips a year to various locations in the tropical Indo-Pacific. In 1999 she retired from her anesthesiology practice and now devotes her time to diving and photographing nudibranchs and other beautiful marine creatures. In 2001 she was appointed a museum associate in malacology at the Natural History Museum of Los Angeles County, where she works under the direction of Angel Valdes, Ph.D.

GERALD R. ALLEN, PH.D., is an internationally renowned authority on the classification and ecology of coral reef fishes of the Indo-Pacific. He was curator of fishes at the Western Australian Museum for twenty-five years, retiring in 1998. He now consults full-time for Conservation International, conducting marine biological surveys, primarily in Southeast Asia. A past president of the Australian Society for Fish Biology, Allen is also an honorary foreign member of the American Society of Ichthyology and Herpetology and the 2003 recipient of the K. Radway Allen Award for Outstanding Contributions to Australian Ichthyological Science. He is the author of thirty-six books and more than four hundred scientific publications. Allen lives in Perth, Australia, where he maintains a rigorous research program from the Western Australian Museum and pursues his avid interest in underwater photography.

STEVEN L. BAILEY came to the New England Aquarium in 1984 while pursuing graduate work in marine biology. Since 1993 he has been the aquarium's curator of fishes, managing the aquarium's collection of birds, reptiles, amphibians, fishes, invertebrates, and plants. The animals under his care range from the fish, turtles, and sharks circling the Giant Ocean Tank to the colony of over eighty penguins inside the aquarium's main entrance to the piranhas and anaconda in the Amazon Rainforest exhibit. He has developed and managed new installations of live animals, both permanent and temporary, including exhibits on sea jellies, frogs and turtles, and the fauna of the Everglades. Bailey is a member of the Association of Zoos and Aquariums (AZA), where he has served on the Field Conservation and Wildlife Conservation Management committees. As co-coordinator of the Species Survival Program for Lake Victoria (East Africa), he is responsible for planning and implementing the AZA's conservation program for the Lake Victoria fishes. A zoologist with a degree from Wilkes University, Bailey is an authority on maintaining live marine fishes in captivity, and he is in demand as a consultant to aquariums the world over.

ROBERT BARREL has owned and operated Nai'a Cruises in Pacific Harbor, Fiji, since 1993. He and his wife, Cat Holloway, first dove in the Phoenix Islands during an expedition to search for flight pioneer Amelia Earhart in 1996, and their discovery of a pristine marine ecosystem has led to four subsequent expeditions there aboard their liveaboard dive boat *Nai'a*.

CAT HOLLOWAY grew up in Australia and learned to scuba dive in Papua New Guinea when she was seventeen. She has been working in or around the dive industry ever since. Before joining Nai'a Cruises as dive master, she was editor of *Scuba Diver Australasia*. Together with her husband, Rob Barrel, she has pioneered boat-based ecotourism in Fiji and Tonga.

SUE MILLER-TAEI is Pacific Islands Marine Director for Conservation International. She has worked on the Phoenix Islands Protected Area since mid-2005, helping to design the marine protected area, draft its management plan, and nominate it for World Heritage site status. She also leads engagement for CI's support of the Pacific Oceanscape. Her previous experience includes consultancy and staff positions with the Pew Environment Group, IUCN, WWF, the International Fund for Animal Welfare, and the South Pacific Regional Environment Programme. She has lived and worked in the South Pacific for more than twenty-five years, designing and managing MPAs and coastal and marine environments, working on advocacy, education, and awareness initiatives, and helping to develop policies governing the use of ocean resources. She is part Samoan Kiwi, an "islander" by heart and nature, with blood and family ties in Samoa, New Zealand, French Polynesia, and Fiji.

EDUARD NIESTEN, PH.D., directs the Conservation Stewards Program at Conservation International (CI), focusing on the use of negotiated agreements with resource owners to achieve conservation and development outcomes. During ten years with CI, he has contributed to field projects in South and Central America, Africa, Asia, and the Pacific as well as research efforts to refine conservation tools and approaches. He holds a doctorate in applied economics from Stanford University.

CHRISTOPHER PALA is an independent scholar and journalist based in Honolulu, Hawaii. His article on the Phoenix Islands Protected Area, "Victory at Sea," appeared in *Smithsonian* in 2008. Pala has worked as a reporter since graduating from the University of Geneva in 1974. He has covered stories in Puerto Rico, the Caribbean, West Africa, Russia, and Central Asia. Pala is the author of *The Oddest Place on Earth: Rediscovering the North Pole*.

RAY PIERCE, PH.D., is a consultant conservation biologist with a long interest in shorebird and seabird biology. His research and management projects have included recovery initiatives for many endangered birds, including New Zealand's black stilt, brown kiwi, and brown teal, as well as the Tuamotu sandpiper of French Polynesia and Phoenix petrel and other seabirds of Kiribati. He has increasingly focused his work on ecological restoration of habitats for these and other birds, including pest eradication initiatives in the Phoenix Islands and Kiritimati of Kiribati. Pierce lives at Speewah in northern Queensland, Australia.

RANDI D. ROTJAN, PH.D., is an associate research scientist at the New England Aquarium. Her tropical coral reef research focuses mainly on fish-coral interactions, but also includes areas such as symbiosis, food webs and trophodynamics, behavioral ecology, and conservation biology. Rotjan sits on the editorial board of the journal *Coral Reefs* and maintains an active global research program with work in the Caribbean, Red Sea, Indo-Pacific, and Central Pacific. She uses an approach that combines field, lab, and computational tools to answer questions about coral reef ecology. Rotjan received her doctorate from Tufts University in 2007, then held a postdoctoral fellowship at Harvard University before joining the New England Aquarium.

PETER SHELLEY is vice president and senior attorney with the Conservation Law Foundation in Boston, Massachusetts. He was awarded a Pew Fellowship in Conservation and the Environment in 1996 and the David B. Stone Medal by the New England Aquarium in 2003. Before joining CLF, Peter served for five years as an assistant attorney general for the Pennsylvania Department of Environmental Resources.

HEATHER TAUSIG is associate vice president of conservation at New England Aquarium. She is responsible for overseeing strategic direction and fund-raising for the aquarium's conservation initiatives, including its sustainable seafood programs, which aim to protect the world's ocean resources by raising public awareness and working with the seafood industry to advance sustainable practices within wild-capture fisheries and aquaculture operations. She currently serves on the advisory boards of EcoFish and the University of Massachusetts's Large Pelagics Research Center, as well as on the Food Marketing Institute's Sustainable Seafood Working Group's Advisory Council. Tausig received her master's degree in international relations and energy and environmental studies from Boston University.

TUKABU TEROROKO is director of the Phoenix Islands Protected Area, Tarawa, Kiribati. He previously served as permanent secretary of the Kiribati Ministry of Environment, Lands and Agriculture Development (MELAD). He has been involved with the PIPA since it was declared by the government of Kiribati in 2006.

INDEX

Page numbers followed by a *p* indicate a photo and by a *t* indicate a table.